Cambridge Elements ≡

Elements in the Philosophy of Biology
edited by
Grant Ramsey
KU Leuven, Belgium
Michael Ruse
Florida State University

ECOLOGICAL COMPLEXITY

Alkistis Elliott-Graves
Bielefeld University

CAMBRIDGE
UNIVERSITY PRESS

CAMBRIDGE
UNIVERSITY PRESS

Shaftesbury Road, Cambridge CB2 8EA, United Kingdom

One Liberty Plaza, 20th Floor, New York, NY 10006, USA

477 Williamstown Road, Port Melbourne, VIC 3207, Australia

314–321, 3rd Floor, Plot 3, Splendor Forum, Jasola District Centre, New Delhi – 110025, India

103 Penang Road, #05–06/07, Visioncrest Commercial, Singapore 238467

Cambridge University Press is part of Cambridge University Press & Assessment, a department of the University of Cambridge.

We share the University's mission to contribute to society through the pursuit of education, learning and research at the highest international levels of excellence.

www.cambridge.org
Information on this title: www.cambridge.org/9781316514122

DOI: 10.1017/9781108900010

First published 2023

A catalogue record for this publication is available from the British Library

ISBN 978-1-316-51412-2 Hardback
ISBN 978-1-108-82752-2 Paperback
ISSN 2515-1126 (online)
ISSN 2515-1118 (print)

Cambridge University Press & Assessment has no responsibility for the persistence or accuracy of URLs for external or third-party internet websites referred to in this publication and does not guarantee that any content on such websites is, or will remain, accurate or appropriate.

Ecological Complexity

Elements in the Philosophy of Biology

DOI: 10.1017/9781108900010
First published online: July 2023

Alkistis Elliott-Graves
Bielefeld University

Author for correspondence: Alkistis Elliott-Graves,
a.elliott-graves@uni-bielefeld.de

Abstract: Complexity has received substantial attention from scientists and philosophers alike. There are numerous, often conflicting, accounts of how complexity should be defined and how it should be measured. Much less attention has been paid to the epistemic implications of complexity, especially in Ecology. How does the complex nature of ecological systems affect ecologists' ability to study them? This Element argues that ecological systems are complex in a rather special way: they are causally heterogeneous. Not only are they made up of many interacting parts, but their behaviour is variable across space or time. Causal heterogeneity is responsible for many of the epistemic difficulties that ecologists face, especially when making generalisations and predictions. Luckily, ecologists have the tools to overcome these difficulties, though these tools have historically been considered suspect by philosophers of science. This is an updated philosophical account with an optimistic outlook of the methods and status of ecological research.

Keywords: complexity, heterogeneity, ecology, generality, prediction

ISBNs: 9781316514122 (HB), 9781108827522 (PB), 9781108900010 (OC)
ISSNs: 2515-1126 (online), 2515-1118 (print)

Contents

1 Introduction   1

2 What Is Ecological Complexity?   6

3 What Are the Effects of Ecological Complexity?   24

4 Dealing with Ecological Complexity   46

5 Concluding Remarks   66

  Bibliography   68

1 Introduction

Why should we care about ecological complexity? This question can be understood in two ways, depending on whether the emphasis is placed on the word 'ecological' or the word 'complexity'. We could be asking why we should care about complexity as it manifests in ecology, rather than how it manifests in other disciplines. Alternatively, we could be asking why, of all the issues in ecology, we should focus on complexity. It turns out that these two questions are closely connected; what makes *complexity* in ecology interesting is also what makes complexity in *ecology* interesting. The short answer is *causal heterogeneity*: the variability of causal factors over space or time. In what follows, I will argue that causal heterogeneity is an important but hitherto undervalued dimension of complexity. It is important because it explains some of the most pressing difficulties faced by practicing ecologists, namely generalisation, prediction and intervention in ecological systems. A re-conceptualisation of complexity that includes causal heterogeneity can give us a better understanding of these problems.

The idea that complexity creates difficulties for scientific practice is neither new nor limited to ecology. It features prominently in the debate about laws in biology, as it explains why most generalisations in biology fall short of the standards of 'lawhood' (Mitchell 2003). Biological systems are complex in the sense that they contain numerous causes whose interactions lead to configurations that are contingent on historical factors. As a result, any generalisations that describe them are neither universal nor exceptionless (Mitchell 2003). This has been a conspicuous thorn in the side of many biologists and philosophers of biology, as historically, laws were considered to be the hallmarks of true science. Any discipline that did not have laws of nature was at best immature and at worst not truly scientific.

While some biologists and philosophers gave up on the idea of biological laws completely (e.g. Lawton 1999, Shrader-Frechette & McCoy 1993)[1], others argued that if laws in biology do not conform to our pre-existing conception of lawhood, then the fault lies with this conception; the answer is to revise our notion of lawhood so that it captures biological laws (Mitchell 2003, Woodward 2001). In the words of Sandra Mitchell, the plurality of causes in evolutionary biology is 'not an embarrassment of an immature science, but the mark of a science of complexity' (2003, p. 115). Here, Mitchell succinctly highlights the two main issues of biological complexity: that complexity is a key way in which

---

[1] Not everyone interpreted the absence of laws in biology as equally problematic. For instance, Shrader-Frechette and McCoy (1993) argued that the absence of laws did diminish the scientific status of biology.

biology differs from other sciences and that this does not mean that biology is deficient or inadequate when compared to these other sciences.

Ecological systems are biological systems and thus share most of the features of biological complexity along with the difficulties it generates. However, ecological systems are special in the sense they are also characterised by pervasive *causal heterogeneity*. Causal heterogeneity exacerbates and compounds the difficulties generated by biological complexity: not only do ecological systems contain numerous causes, but these causes are also diverse and variable. This means that even the generalisations that are present in evolutionary biology might be elusive in ecology. Thus, a thorough investigation of *ecological complexity*, with causal heterogeneity as one of its key features, is important for gaining a deeper understanding of some of the most important problems faced by practicing ecologists. In addition, it can help us gain a more comprehensive understanding of scientific practice, as understanding ecological complexity can serve as a blueprint for a better understanding of complexity in other disciplines where causal heterogeneity also features prominently.

But how important is ecological complexity *really*? Is rarity or absence of laws merely a philosophical problem or does it also affect the practice of ecology? The effects of complexity are far reaching for ecological practice *and* for the theoretical foundations of the discipline. I will illustrate the practical effects of complexity in the next section, by showing that complexity can lead to *surprises* in ecological research. I will then show that frequent surprises have dangerous theoretical implications, as they are used by some scientists and philosophers to cast doubt on the overall quality of ecological research and the scientific status of the discipline itself.

1.1 Surprise!

While investigating the effects of bird guano runoff on intertidal ecosystems in southwestern South Africa, a group of scientists observed that two neighbouring islands (4 kilometres apart) had very different benthic communities: one was teeming with lobsters while the other was covered in mussels and whelks. According to the local fishermen, lobsters were present in both locations till the early 1970s, but then mysteriously disappeared from the second island. After a series of horror-inducing experiments, where lobsters were re-introduced to the second island, the scientists realised that the whelks had turned the tables on their erstwhile predators and now preyed on the lobsters (Barkai & McQuaid, 1988).[2] Another example of surprise comes from a species

---

[2] The horrifying aspect was the speed with which the whelks consumed their erstwhile predators: about 1,000 lobsters were completely annihilated within 45 minutes (Wilcox, 2018).

of butterfly that lays its eggs in a particular host plant (Singer & Parmesan, 2018). An invasive plant outcompeted the butterfly's native host, but the butterfly adapted to the invader. Around twenty years later, the invader was eradicated, but the butterfly could not switch back to its original host and went locally extinct. A third example comes from interactions between plants and soil microbes, which have been known to reverse, changing from positive to negative feedback (Casper & Castelli, 2007; Klironomos, 2002; see also Section 2.2). Finally, Benincà et al. (2008) observed unexpected and significant changes in species abundances and community structure in an experiment on a plankton community isolated from the Baltic Sea, even though the experiment was conducted in a controlled laboratory setting with most conditions kept constant.

The frequency of surprises like these in ecology has been documented. Doak and colleagues (2008) conducted a survey and found that surprises are far from rare. They outlined at least sixteen cases of famous surprises just within the subfields of population and community dynamics and reported that 98 per cent of established field ecologists affirmed that they had encountered surprise events akin to the sixteen cases. Moreover, many of the respondents revealed that the majority of surprising results had not been subsequently sent for publication, 'the implication being that these observations were uninteresting, bothersome, embarrassing, or not sufficiently well chronicled and understood through proper application of the scientific method, and thus were *underreported* in the scientific literature' (p. 956, my emphasis).

But what does the existence and frequency of surprises mean for ecological research? In the philosophical literature, a few surprises are viewed as a positive and integral aspect of scientific practice. Scientists learn from surprises, as understanding why they occur leads to scientific progress (Morgan, 2005; Parke, 2014). However, too many surprises are problematic. In the above examples, the scientists had identified patterns in nature (or the laboratory) and formulated expectations based on those patterns. Yet these patterns were ephemeral: they existed for a while, but at some point, they ceased, resulting in surprise. This explains why surprises are problematic: scientists rely on identifying patterns to generate generalisations, on which they base explanations, predictions and interventions. A surprise indicates that the explanation, prediction or intervention has failed or is likely to fail.

Ecological complexity (which includes the notion of causal heterogeneity) explains both the frequency and magnitude of surprises in ecology. Ecological phenomena have numerous, diverse and variable causes, so the behaviour of ecological systems does not always go as expected. This diversity and variability is the reason why the patterns that scientists detect are likely to be ephemeral, resulting in surprise. Nonetheless, as we shall see in the next section, there is a view which states that frequent surprises only occur in disciplines that are immature or whose theories or methods are somehow flawed (Doak et al., 2008; Hitchcock & Sober, 2004).

1.2 The Scientific Status of Ecology

Some ecologists believe that the frequency of surprises in their discipline should be taken as an indication that there are gaps in our knowledge of ecological systems. In some sense, worrying about the quality of a discipline's research is something that all scientists (ought to) do, as it helps to maintain standards and improve methods in scientific practice (Hitchcock & Sober 2004).[3] However, there are ecologists who go much further, questioning whether ecology should be considered a science at all, or lamenting that it is at best a 'soft' science. Ecologists are often said to suffer from 'physics envy', wishing that their theories, methods and results would more closely emulate those in physics (Egler, 1986; Kingsland, 1995 p. 234; McIntosh, 1987; Shrader-Frechette & McCoy, 1993 p. 34). Another phrase sometimes invoked is that of 'stamp collecting', which is the lot of scientific endeavours that are merely descriptive, lacking general theories, predictive power and the ability to be expressed mathematically (Johnson, 2007; Kingsland, 1995 p. 200). As historian Sharon Kingsland points out, the introduction of mathematical models into ecology was viewed by ecologists themselves as an important step towards the discipline becoming a real science, and that this trend is ongoing, as 'ecologists continue to look towards mathematics and the physical sciences for ideas, techniques and models of what science should be' (1995, p. 234).

Despite these 'advances', there is a small but persistent and vocal group of ecologists who continue to worry. Every few years publications appear, often in monographs or the opinion section of major journals, expressing misgivings about the scientific status of ecology or one of its sub-disciplines. Perhaps the most famous of these critiques is Peters's aptly titled *Critique for Ecology* (1991), which criticised ecologists for not providing testable hypotheses in the form of precise predictions. Moreover, Peters argued that theory did not play a significant enough role in ecological research as it did little more than provide the conceptual inspiration for a scientific investigation. It seems that many ecologists took this criticism to heart, as 'across the western world there were professors who removed the book from library shelves to prevent their students from reading it, lest they became demotivated' (Grace, 2019). A more recent version of this view appears in Marquet et al. (2014), who argue that ecology does not have enough 'efficient theories', by which they mean theories

---

[3] I thank Jack Justus for pointing this out.

that 'are grounded in first principles, are usually expressed in the language of mathematics, make few assumptions and generate a large number of predictions' (p. 701). On a similar note, Houlahan et al. (2017), argue that ecology has 'abandoned prediction [and] therefore the ability to demonstrate understanding' (p. 1). Moreover, it is still an 'immature discipline . . . [that] must move beyond such qualitative coarse predictions to riskier, more quantitative, precise predictions, sensu Popper'(p. 5).

There are also critiques of sub-disciplines or types of research, which are in some sense more alarming, as they could be used by universities or funding bodies to limit the amounts allocated to those disciplines or methods. For example, Valéry et al. (2013) argued that their inability to find a process or mechanism specific to invasion biology 'eliminates any justification for the autonomy of invasion biology' (p. 1145). Courchamp et al. (2015) state that 'one of the central objectives and achievements of fundamental ecology is to develop and test general theory in ecology' (p. 9). Here, fundamental, or 'pure' ecology is contrasted to 'applied' ecology, which is aimed at solving particular problems and/or intervening on the world. The authors worry that applied ecology has seen an increase in support (economic and otherwise) in recent years, at the expense of fundamental ecology, which should be reversed (Courchamp et al., 2015).

Though it is primarily ecologists who worry about the scientific status of their field, these ideas are rooted in philosophy of science. As stated above, the ability to generate laws and the ability to make precise and accurate predictions used to be seen as *the* hallmarks of true scientific disciplines (Hempel & Oppenheim, 1948; W. C. Salmon, 2006). A discipline that could not provide either, would traditionally be considered at best 'immature' and at worst 'soft' or 'unscientific' (Rosenberg, 1989; see discussion in Winther, 2011). The more extreme versions of the positions are nowadays viewed as outdated in philosophical circles, yet aspects of them are still deemed important. Many philosophers arguing for a revised notion of laws, do so partly in order to show that sciences like biology are on a par with other sciences. For example, Linquist et al. (2016) argue that as ecology has resilient generalisations which ought to count as laws, this 'should help to establish community ecology as a generality-seeking science as opposed to a science of case studies' (p. 119).

Thus, there seems to be a general worry that ecology is far from an ideal science. The suggestions for how to improve the quality of ecological research vary: some argue that the answer is to find more or better laws, others argue for more focus on explanations or predictions, others still argue for more integration between sub-disciplines, and so on. My view is different, as I do not believe that there is anything, in principle, wrong with ecological research, merely that ecological research is particularly difficult in certain ways. These difficulties stem from the particular way in which ecological systems are complex.

1.3 Outline

My aims, in this Element, are to show that (i) our current views about complexity do not capture how complexity works in ecological systems (ii) we should reconceptualise complexity to include causal heterogeneity (iii) this reconceptualisation explains some of the important difficulties that ecologists face and (iv) this reconceptualisation can point to some ways of mitigating these difficulties. The argument will proceed as follows. In Section 2, I examine the concept of 'complexity' in ecology. I start by providing a brief sketch of the main characteristics associated with complexity and then move to an in-depth account of some of these characteristics, that is, those that affect the study of ecological systems. I then turn to Levins's (1966) account of complexity and trade-offs between model desiderata, which I subsequently extend and refine. In Section 3, I examine some of these trade-offs in more detail, showing how causal heterogeneity creates difficulties for generalisations, predictions and interventions. In Section 4, I argue that this explains but does not justify the worry that ecology is not a true or sufficiently mature science. I show that even if we give up on extensive generalisation in ecology, ecologists are capable of making successful predictions and interventions. Rather than being embarrassed by the modesty of ecological generalisations, ecologists and philosophers should recognise the scientific and practical value of ecology's methodological toolkit. In Section 5, I outline some concluding remarks on generalisation and prediction in science more broadly.

This Element is not just meant for a philosophical audience. I hope that any ecologists looking for an alternative philosophical view of science, that accounts for the peculiarities and idiosyncrasies of their discipline, will find the arguments I present helpful. Moreover, I hope that this philosophical approach can be used by practicing scientists to support the alternative, undervalued research strategies examined in Section 4. Finally, this discussion of ecological complexity could also be helpful for scientists in other disciplines whose systems are also causally heterogeneous, such as Economics or Climate Science.

2 What Is Ecological Complexity?

The claim that ecological systems are complex is uncontroversial. Simon Levin's declaration that ecosystems are 'prototypical examples of complex adaptive systems' (Levin, 1998) is frequently taken as the starting point for discussions of complexity in ecology (Parrott, 2010; Proctor & Larson, 2005; Storch & Gaston, 2004). Nonetheless, there is no simple answer to the question, 'what is ecological complexity?' as there is no single, universally accepted definition of ecological complexity. Instead, there are a number of diverse and not always overlapping characterisations originating in various disciplines, including biology, physics and social science (Bascompte & Solé, 1995; Donohue et al., 2016; Levin, 2005; Shrader-Frechette & McCoy, 1993; Storch & Gaston, 2004).

As there is no short answer to the complexity question and no comprehensive definition of the term, the aim of this section is to provide a guide for thinking about the question 'what makes an ecological system complex?' I will start with some background on the concept of complexity, as it was discussed within and outside ecology (Section 2.1). In Section 2.2, I will outline the most important characteristics of ecological complexity. In Section 2.3. I will examine the epistemic[4] implications of complexity, namely difficulties in generalising, predicting and intervening on ecological systems. In Sections 2.4 and 2.5, I will connect the discussions of the previous sections, arguing that in the context of epistemic difficulties, ecological complexity should be understood as the combination of 'having multiple parts', 'interaction' and 'causal heterogeneity'.

2.1 Ecological Complexity in Context

Though aspects of complexity have been studied in various disciplines for more than 150 years, interest in complex systems began in earnest in the 1960s and 1970s, and gained momentum in the 1980s (Hooker, 2011; Miller & Page, 2009; Simon, 1962; Wimsatt, 1972). The subsequent explosion of research on complexity and its effects on a variety of phenomena, had important and long-lasting implications for scientific practice, as it contributed to the establishment of a framework for anti-reductionist philosophy of science (Hooker, 2011; Mitchell, 2009), along with the recognition that the emergence and manifestation of complexity, especially in biological systems, is an worthwhile and fruitful research topic (McShea & Brandon, 2010; Mitchell, 2009; Wimsatt, 1972).

Despite – some might say because of – the level of interest and research in complex systems, a single, unified definition of complexity has yet to be agreed on (Hooker, 2011; Ladyman et al. 2013; Miller & Page, 2009). In lieu of a precise or formal definition of complexity, scientists and philosophers usually list some characteristics that tend to appear in complex systems. It is worth noting that there

---

[4] For readers without a background in philosophy, the term 'epistemic' here means related to *knowledge*. I am interested in the effects of ecological complexity on what ecologists know about the systems they investigate, how they know it and what difficulties arise in the acquisition of this knowledge.

is quite a bit of overlap, as some characteristics appear frequently. One such characteristic is having 'multiple interacting parts', which seems to be a basic requirement of complexity. The idea is that a high number of parts and (dynamic) interactions between them increase the likelihood of complex behaviours. Table 1 contains a representative selection of characterisations of complexity, from complex systems science and (philosophy of) biology[5]. As we can see, multiple interacting parts (highlighted in bold), features prominently.

Within the discipline of ecology, the context of complexity has its own history. Here, discussions of complexity were originally related to a question that was considered fundamental, namely, 'how are ecological systems possible?' Early influential ecologists, such as Odum, Elton and MacArthur, found the ability of populations, communities and ecosystems to persist, in spite of internal and external disturbances, quite remarkable, hypothesising that this apparent stability was caused by the diversity and connectivity of ecological communities (Kingsland, 2005; McCann, 2000 Odenbaugh, 2011). The general idea is that complex communities are more able to adapt to changes, such as disturbances or perturbations (fires, new competitors/predators, sudden climatic changes), without falling apart. This view was famously disputed by Robert May (1973), who used mathematical models to show that we should expect complex communities to be less stable. Subsequent generations of ecologists have refined the concepts of diversity, complexity and stability in order to bolster their favoured side of the debate, while philosophers of ecology have provided their own clarifications and categorisations of the various views (McCann, 2000; Odenbaugh, 2011). A current consensus seems to be that complexity is indeed an inherent feature of healthy and mature ecological systems, even though such systems may be susceptible to particular disturbances (Hooper et al., 2005; Loreau et al., 2001; McCann, 2000; Parrott, 2010).

The brief outline of the context of ecological complexity highlights two important points for our discussion. First, it is uncontroversial, indeed quite common to consider multiple interacting parts as key features of complex systems, including biological systems. Thus, there is also no difficulty in recognising that it is also a key feature of complex ecological systems. Second, whether or not ecologists agree that complexity leads to stability, they seem to agree that complexity is an inherent feature of (at least healthy) ecosystems. This is an important theme that will appear throughout the Element: the complexity of ecological systems is an inherent feature of the systems themselves. In other words, it is inescapable.

---

[5] The first five quotes on the table have been taken from a list in Ladyman et al (2013), who collected quotations from a 1999 special issue in *Science* on complex systems.

Table 1 Characterisations of complexity

Reference	Characterisation
(Whitesides and Ismagilov 1999, p. 89)	a complex system is one whose evolution is very sensitive to initial conditions or to small perturbations, one in which the number of **independent interacting components is large**, or one in which there are multiple pathways by which the system can evolve. Analytical descriptions of such systems typically require non-linear differential equations.
(Weng et al. 1999, p. 92)	the adjective 'complex' describes a system or component that by design or function or both is difficult to understand and verify. [. . .] complexity is determined by such factors as **the number of components and the intricacy of the interfaces between them**, the number and intricacy of conditional branches, the degree of nesting, and the types of data structures.
(Parrish and Edelstein-Keshet 1999, p. 99)	Complexity theory indicates that **large populations of units** can self-organize into aggregations that generate pattern, store information, and engage in collective decision-making.
(Rind 1999, p. 105)	A complex system is literally one in which **there are multiple interactions between many different components**.
(Brian Arthur 1999, p. 107)	Common to all studies on complexity are systems with **multiple elements** adapting or reacting to the pattern these elements create.
(Simon 1962, p. 468)	I shall not undertake a formal definition of 'complex systems'. Roughly, by a complex system I mean one made up of **a large number of parts that interact** in a nonsimple way. In such systems the whole is more than the sum of the parts, not in an ultimate, metaphysical sense but in the important pragmatic sense that, given the properties of the parts and the laws of their interaction, it is not a trivial matter to infer the properties of the whole.
(Tëmkin 2021, p. 299)	The complexity of biological dynamics stems from the synergetic effect of idiosyncratic processes at different organisational levels and the **dynamics of interlevel interactions**

2.2 Key Features of Ecological Complexity

Now that we have a rough idea of how scientists and philosophers characterise complexity, we can take a closer look at the most important characteristics or features that are usually associated with complexity. This is not meant to be an exhaustive list of all the features ever to be associated with complexity, but a summary of the characteristics that are relevant for the context of ecological research, namely, *multiple parts, diversity, interaction, emergence, non-linearity* and *historicity.*

Multiple Parts. Complex systems tend to be composed of many parts. Increasing the number of parts allows for more and varied interactions between them, which facilitates complexity. Even though this characteristic is frequently included in characterisations of complexity, it is usually not discussed in great detail (especially outside biology), perhaps because it seems conceptually straightforward. An example of this characteristic 'in action' can be seen in McShea and Brandon's (2010) account of complexity in evolutionary biology. Here, a high or increasing number of parts is taken as a marker of increase in complexity, which in turn is associated with evolutionary progress. The basic idea is that, in the absence of other constraints, evolution creates organisms of increasing complexity. Even if some species are less complex than their ancestors, the claim is that complexity increases over evolutionary time.

Diversity (aka heterogeneity). There are various notions of diversity in ecology, the most famous of these being *bio*diversity (Justus, 2021; Levin, 2002; Maclaurin & Sterelny, 2008; Parrott, 2010; Santana, 2014). Biodiversity is a notoriously difficult concept to define and measure (Maclaurin & Sterelny 2008) so much so that some have questioned the usefulness of the concept for ecological research (Santana 2014, see also discussion in Justus 2021, section 5). I will not delve into the debates surrounding the definition, measurement and value of biodiversity. To avoid entering the territory of these debates, I will follow Levins (1966) and Matthewson (2011) in focusing on *heterogeneity*: the diversity between parts of a system or between systems. For example, a population can be heterogeneous because its individuals are diverse in terms of genetic and behavioural traits, a community can be heterogeneous when it is composed of populations of different species, while an ecosystem can be heterogeneous when it is composed of multiple communities. How exactly heterogeneity should be understood and how it relates to other characteristics of complexity is a central theme in this Element. I worry that when heterogeneity is subsumed under the notion of multiple parts (see for example Levin, 2002; McShea & Brandon, 2010; Odenbaugh, 2011; Parrott, 2010; Potochnik, 2017;

Wimsatt, 1972), it is too easily overlooked. In Sections 2.2 and 2.3 I examine my view of heterogeneity and its relationship to the characteristic of 'multiple parts'.

Interaction. Another important characteristic of complex systems is that the parts of the system *interact* with each other. That is, the parts of a system stand in cause-and-effect relationships. These effects are often (but need not be) non-linear or non-additive. Outside ecology interaction is not always treated as an independent characteristic, but often subsumed under one of the others, such as multiple parts, feedback, non-linearity or emergence (Hooker, 2011 but see Ladyman et al., 2013). Within ecology, interaction of parts features more prominently as a characteristic in its own right. For example, Stuart Pimm's influential account of complexity included interaction as part of its definition, which he further categorised into *connectance*, that is, the number of interspecific interactions out of those possible, and *interaction strength*, that is the average value of the interspecific interactions in the community (Odenbaugh, 2011).

Emergence. This refers to the phenomenon of lower-level parts within a system giving rise to higher-level behaviours. For example, *density dependence* is the effect that the size of a population has on its members. Individual members of the population cannot display density; it is a property of the population as a whole. Discussions of emergence often have ideological connotations, in the sense that they imply a deeply anti-reductionist basis for complex systems research (Mitchell, 2009). The idea is that even though systems are made up of parts, it is (sometimes, often or always – depending on the strength of the view) better to investigate the behaviour of the system at the higher level – the level of the emergent property, rather than at the level of the constituent parts. What exactly is meant by 'better' here also depends on the view of emergence, but usually refers to some aspect of explicability: it is easier, more understandable, or more comprehensive to provide an explanation that includes the higher-level property, rather than one that occurs just at the lower level. Some emergent properties are considered to be representative characteristics of complex systems. For example, complex systems are organised *hierarchically*, that is, their components are grouped at different levels, and higher-level groups constrain the behaviour of lower groups. This contributes to *self-organisation* and *causal autonomy* that is, the ability of the system to regulate its own states and/or behaviour, creating and maintaining the processes that enable it to function (Levin, 2002, 2005). An example of a biological hierarchical, self-organised and autonomous system is an organism's metabolism, which creates and maintains the processes enabling the organism to live (Hooker, 2011).

Non-linearity. A system is linear if 'one can add any two solutions to the equations that describe it and obtain another, and multiply any solution by any factor and obtain another' (Ladyman et al., 2013, p. 36). Complex systems do not possess this quality hence they are *non-linear*. In linear systems, a change in one factor will result in a similar or at least proportional change in the behaviour of the system. In non-linear systems, a change in one factor can result in a disproportional change in the behaviour of the system. For example, the *growth rate* of populations is positive when these populations are small and resources are abundant, yet as the population grows it does not do so proportionally. If resources continue to be abundant, the growth rate itself increases, yet if they become scarce, the growth rate becomes negative (i.e. the population decreases).

Non-linearity is sometimes linked with other characteristics of complexity, most notably *chaos* or *sensitivity to initial conditions* (Anand & Orlóci, 1996; Benincà et al., 2008; Ladyman et al., 2013). It is often difficult to distinguish between these concepts and their definitions: sometimes they are used interchangeably, sometimes one is considered to be a characteristic of another. For example, Bishop (2011) defines sensitivity to initial conditions as 'the property of a dynamical system to show possibly extremely different behavior with only the slightest of changes in initial conditions' (p. 108), and characterises the property as a feature of both chaos and complexity.

Another characteristic that is sometimes linked to non-linearity is *feedback*. Feedback can greatly magnify an initial (small) effect, contributing to the non-linearity of the system. For example, removing a keystone species from an ecosystem can result fundamental changes in the ecosystem, including a cascade of local extinctions and the eventual collapse of the entire ecosystem (Levin, 1998).

Path dependence (aka historicity). Put simply, systems display this characteristic when their later states depend on their previous states. Sometimes, development along a certain path becomes increasingly entrenched. For example, an initial mutation that provides a competitive advantage then spreads through the population (Hooker 2011, p. 33). An important ecological version of path dependence is that of Simon Levin. Here, path dependence is an effect of non-linearity, as it occurs when 'the local rules of interaction change as the system evolves and develops' (Levin 1998, p. 433). A typically ecological example of path dependence is the colonisation of new areas, such as islands or forest patches. The final composition of the system (i.e. which species persist and in what proportion) will depend on which species are the original colonisers and how they interacted with each other.

2.3 The Effects of Complexity

So far, I have given a brief outline of what complexity is, in terms of some key characteristics. The aim of this Element is to examine the characteristics that are relevant for a particular context: what difficulties complexity creates for ecological research. The starting point for this discussion is the work of Richard Levins, specifically his (1966) paper 'The structure of model building in population biology', which, as we shall see, explicitly focuses on the effects of complexity for biological research. Levins's argument was subsequently refined (Levins 1993), and his ideas have received attention in philosophy of science, especially in the literature on modelling (Godfrey-Smith, 2006; Justus, 2005, 2006; Matthewson, 2011; Odenbaugh, 2003; Orzack, 2005; Orzack & Sober, 1993; Weisberg, 2006). This body of philosophical literature forms the conceptual background for many of the ideas in this Element and I will be drawing from it heavily, especially for the remainder of this section.

Levins starts by pointing out that population biologists must deal with systems that are incredibly complex in the sense that: (i) they are made up of many different species, (ii) there is genetic, physiological and age diversity within each population, (iii) there are demographic interactions within and between populations and (iv) all this is happening in heterogeneous environments (p. 421). He recognises that dealing with such a complex system is problematic, as capturing all this complexity in a mathematical model is highly impractical. Even if it were feasible, it would yield results that made little sense to us (Levins, 1966). Thus, scientists must decide how much complexity to include in each model. The issue is that there is no universal optimal level of model complexity, as scientists use models for multiple distinct purposes (i.e. understanding, predicting and modifying nature) (p. 422). The optimal level of model complexity can thus differ, depending on the model's intended purpose and the system it is applied to.

Much of the complexity of ecological systems is represented within models by what Levins calls *realism*: how accurately a model captures the causal structure of the world. In practice, this is achieved by models containing many variables. That is, models are realistic when they are not overly simplified or idealised, for example, when they include many variables that correspond to real world factors, represent (dynamical) links between these variables and relax simplifying assumptions such as symmetry (Levins 1993, pp. 548–52). In contrast, examples of highly unrealistic models in biology include those that 'omit time-lags, physiological states, and the effect of a species' population density on its own rate of increase' and contain assumptions analogous to 'frictionless systems or perfect gasses' (1966, p. 422). I should note that Levins's notion of realism is relevant in current ecological literature, as the term 'model complexity' is still used to refer to the number of variables represented in a model (Clark et al., 2019; Ward et al., 2014).

Capturing system complexity is not the only requirement of a model. Levins identifies two other 'desiderata' that scientists aim for in their models. The second desideratum is *generality*. Put simply, a model is general when it applies to many

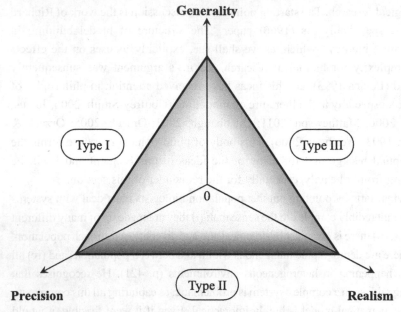

Figure 1 Levins's 3-way trade-off

systems in the world, that is, when a modeller can use the model to gain information about many systems (I explain model applicability in the next two paragraphs). The third desideratum is *precision,* that is, how finely specified a model's predictions are (see footnote 5, Box 1 C and Sections 3.2.1 and 4.3 for a full explanation of precision). On Levins's view (1966, 1993), modellers aim to increase all three desiderata as much as possible within each model. However, after a certain point, only two out of the three desiderata can be increased further. This gives rise to three strategies for building models and three corresponding types of models (Figure 1).

Before I move on to the three types of models, a note on 'model applicability'. In the previous paragraph, I pointed out that models are general when they apply to many systems, which means that they provide information about those systems. This is an intentionally weak constraint, as Levins wants to distinguish between a model merely applying to a system and applying *well* to a system. In principle, anything can be a model for any phenomenon or system. For example, my coffee cup and water bottle could be a model of the earth and the sun. This view may seem controversial, at first glance, but it is espoused by many philosophers of science, who believe that a model's use is partly determined by the intentions of the modeller (see for example Giere 2004; Knuuttila & Loettgers, 2016a; Weisberg, 2013). The weakness of the constraint has an additional benefit, namely that it gives scientists the freedom to use models in novel and creative ways. For example, it allows scientists to use models developed for quite different systems, such as using physical or chemical models in biology or biological models in economics (Knuuttila & Loettgers, 2016b; Weisberg, 2007).

Of course, just because a model *could* be applied to a system, does not mean that it *should* be applied to that system. This depends on the model, the system and the phenomenon we want to investigate. For example, if I want to use my beverage holder model to explain rudimentary planetary motion to my five-year-old nephew, then the model will probably apply quite well to the system. If, on the other hand, I want to use the beverage holder model to predict the next lunar eclipse, then it will be woefully inadequate. There are many accounts in the philosophical literature detailing what it means for a model to *apply well* to a system (Frigg, 2009; Morgan & Morisson, 1999; van Fraassen, 2008; Weisberg, 2013) but adjudicating between these views is not relevant for this discussion. For now, it is more important to be able to identify when a model has been applied well to a system. A common view in the scientific literature, shared by Levins, is that we can *test* that a model applies well to a system, by whether it yields accurate predictions about the system's behaviour (Beckage et al., 2011; Dambacher et al., 2003; Kulmatiski et al., 2011; Stillman et al., 2015; Stockwell, 1999; Tompkins & Veltman, 2006).[6]

Back to the three types of models. Type I models (perhaps the most common models in population ecology) maximise generality and precision at the expense of realism. The main advantage of these models is their ability to apply widely. By omitting all the causal factors specific to particular systems, these models can identify factors that are common across many systems. For example, the logistic model of population growth shows how populations grow when they are limited by the carrying capacity of the environment (see Box 1A). This dynamic is thought to be the core factor of population growth, present in all populations, even when other factors are also present. Thus, the model that describes the process common to all populations is general. Moreover, these models allow scientists to make precise predictions about the future size of a population given a set of initial conditions. The main disadvantage of type I models is that they cannot capture many of the relevant factors and dynamics that are idiosyncratic (i.e. unique to particular systems), so their predictions are often inaccurate (Justus, 2005; Novak et al., 2011). In other words, even though they apply to many systems, they frequently do not apply *well* to these systems.

Type II models (e.g. ecosystem network models) maximise precision and realism at the expense of generality (Box 1B). They are constructed with a particular system in mind, and contain many factors present in the real-world system. The main advantage of these models is that they can capture

---

[6] For a more extensive discussion on the relationship between generality and model applicability, see Elliott-Graves (2022).

BOX 1. TYPE I, II, III MODELS

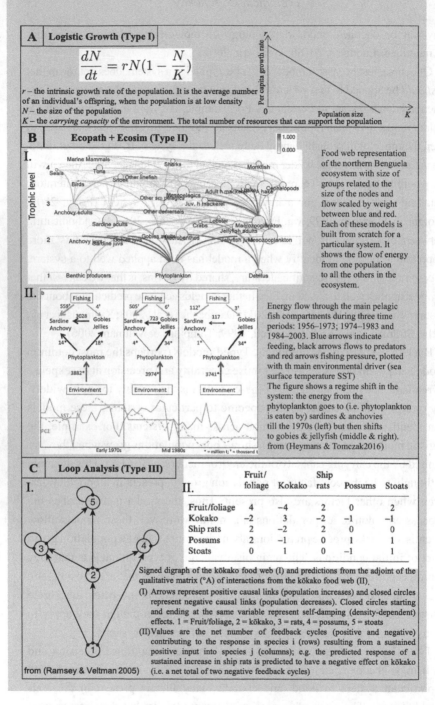

A Logistic Growth (Type I)

$$\frac{dN}{dt} = rN\left(1 - \frac{N}{K}\right)$$

r – the intrinsic growth rate of the population. It is the average number of an individual's offspring, when the population is at low density
N – the size of the population
K – the *carrying capacity* of the environment. The total number of resources that can support the population

B Ecopath + Ecosim (Type II)

I.

Food web representation of the northern Benguela ecosystem with size of groups related to the size of the nodes and flow scaled by weight between blue and red. Each of these models is built from scratch for a particular system. It shows the flow of energy from one population to all the others in the ecosystem.

II.

Energy flow through the main pelagic fish compartments during three time periods: 1956–1973; 1974–1983 and 1984–2003. Blue arrows indicate feeding, black arrows flows to predators and red arrows fishing pressure, plotted with th main environmental driver (sea surface temperature SST)
The figure shows a regime shift in the system: the energy from the phytoplankton goes to (i.e. phytoplankton is eaten by) sardines & anchovies till the 1970s (left) but then shifts to gobies & jellyfish (middle & right). from (Heymans & Tomczak 2016)

C Loop Analysis (Type III)

I.

II.

	Fruit/foliage	Kokako	Ship rats	Possums	Stoats
Fruit/foliage	4	−4	2	0	2
Kokako	−2	3	−2	−1	−1
Ship rats	2	−2	2	0	0
Possums	2	−1	0	1	1
Stoats	0	1	0	−1	1

Signed digraph of the kōkako food web (I) and predictions from the adjoint of the qualitative matrix (°A) of interactions from the kōkako food web (II).
(I) Arrows represent positive causal links (population increases) and closed circles represent negative causal links (population decreases). Closed circles starting and ending at the same variable represent self-damping (density-dependent) effects. 1 = Fruit/foliage, 2 = kōkako, 3 = rats, 4 = possums, 5 = stoats
(II) Values are the net number of feedback cycles (positive and negative) contributing to the response in species i (rows) resulting from a sustained positive input into species j (columns); e.g. the predicted response of a sustained increase in ship rats is predicted to have a negative effect on kōkako (i.e. a net total of two negative feedback cycles)
from (Ramsey & Veltman 2005)

many more factors and dynamics and thus have the potential to provide much more accurate predictions (Fischer et al., 2018; Phillips et al., 2016; Travis et al., 2014). The flip side of the coin is that by including many system-specific factors and dynamics, these models only apply *well* to the systems they are created for; they do not allow scientists to generalize their results beyond the system under investigation (Wenger & Olden, 2012). Another worry for type II models is that they do not always fulfil their potential in terms of predictive accuracy (Schindler & Hilborn, 2015; Ward et al., 2014). This is a version of the problem of 'overfitting', where a complex model includes or accommodates noisy data resulting in inaccurate predictions (Hitchcock & Sober, 2004). The quality of available data is often low in ecology, as it can be partial, gappy or even false – this is referred to as noise (see Section 3.3 *paucity of data*). Type II models, which can incorporate many variables and data points can be 'led astray' by low quality data. They can end up being more susceptible to low quality data than type I models, simply because they can incorporate more of it. Thus, type II models can end up yielding predictions that are highly inaccurate (but see discussion in 4.2.1).

Type III models (e.g. loop analysis) maximise realism and generality but sacrifice precision (Box 1 C). The outputs of these models (their predictions) are not finely specified. They can be imprecise probabilities (e.g. the population will rise by more than 2 per cent), ranges of values (e.g. the time to extinction is 47–60 years) or qualitative trends (e.g. the effect of population A on population B is negative). Their main advantage is that they can maximise both realism and generality simultaneously. They can capture all the relevant variables and dynamics in a system and can also apply to many diverse systems. A second advantage is that they are less susceptible to errors (Justus, 2006). Unlike type I models, they can include many more parameters, thus capturing all relevant dynamics. At the same time, unlike type II models, they are better at avoiding or at least minimising the effect of errors in parameter estimation (because their parameters and outputs are imprecise). However, this way of minimising errors (by widening what can be accepted as accurate) is often seen as a major disadvantage of type III models (Orzack & Sober 1993). The worry is that the imprecise predictions could be masking systematic errors in the models, so that even if the predictions are accurate, they cannot be relied on as tests of how well the model applies to a system, nor can the models be relied on to determine how best to intervene on the systems under investigation (Orzack & Sober, 1993; but see Section 4.3, Elliott-Graves 2020a and Justus 2005, 2006).[7]

Levins's trade-off framework can be understood both as a *diagnosis* of the effects of complexity for scientific practice and as a *suggestion* for how best to

---

[7] Note that precision is not the same as accuracy. Precision refers to how finely specified a result is. Thus, the statement 'the population is > 200 individuals' is less precise than 'the population is 247 individuals'. Both these results could be accurate or inaccurate depending on what the actual size of the population is. In fact, the more precise a result, the likelier it is to be *in*accurate. A more extensive discussion of the relationship between precision and accuracy can be found in Section 3.2.1.

deal with these effects. The diagnosis is that complexity creates difficulties for scientific practice: it imposes limits on how scientists can investigate natural systems. An important assumption of this diagnosis is that complexity is not an artefact of our scientific methods, but a feature of biological systems. This means that scientists have no control over complexity itself (e.g. they cannot reduce the complexity of the systems they study, no matter how much science progresses). They can only get better at dealing with its effects. But this leads to another question: Are the *trade-offs* also caused by complexity or are *they* artefacts of our methods and/or our epistemic limitations (Matthewson, 2011)? If the trade-offs are mere artefacts, then there is no reason to adopt Levins's suggestions for how to deal with complexity.

Levins stated (and subsequent commentators have explicated) that this is a false dichotomy (Levins, 1966; Matthewson, 2011; Weisberg, 2006). Trade-offs exist because of the combination of biological complexity and our epistemic limitations. Complexity ensures that each system has hundreds of parameters, and multiple dynamic interactions between them. Our epistemic and technological limitations mean that the collection of data is difficult and prone to introduction of error. Even if all this data was collected and put in a model, the model would probably be analytically insoluble. Even if we managed to create a model that provided analytical solutions, these solutions would be meaningless to us (Levins 1966, p. 421).[8] In other words, even if part of the problem is that we are limited beings and cannot adequately capture complexity in our models, these limitations are not temporary (Matthewson 2011; Weisberg 2006). Technological advances are likely to make data collection easier or less biased or even result in models that are *better* at capturing complexity. Yet there are limits to these advances so we will never be able to fully capture the complexity of natural systems. It therefore behoves us to adopt scientific strategies that best *deal with the effects* of complexity. I address this in Section 4. In the meantime, I return to the discussion of how we should understand ecological complexity.

2.4 Which Characteristics of Complexity Give Rise to Trade-offs?

In the previous section, I outlined the contextual backdrop which frames our discussion of complexity, namely the effects of complexity on model construction. We can now fine-tune our characterisation of ecological complexity in light of this contextual backdrop. From Section 2.1, we have the following characterisation of complexity (à la Levins): 'Population biology must deal simultaneously with genetic, physiological and age heterogeneity within

---

[8] Weisberg (2006) goes through the implications of each of these points for Levins's framework in detail (pp. 627–33).

species of multispecies systems changing demographically and evolving under the fluctuating influences of other species in a heterogeneous environment. The problem is how to deal with such a complex system' (1966, p. 421). I identified the following characteristics from this description: complex systems in biology are (i) made up of many different species, (ii) there is genetic, physiological and age diversity within each population, (iii) there are demographic interactions . within and between populations and (iv) all this is happening in heterogeneous environments (p. 421). If we cross-reference this description with our list from Section 2.1, we can detect three features that together give rise to complexity: *multiple parts*, *interaction* and *heterogeneity*.

Levins, like many other scientists and philosophers, believes that complex systems are made up of many parts. But what counts as a system and what as a part? Levins identifies individuals, populations and species and bits or aspects of environments as possible 'parts'. This is intentional: a scientist can categorise the world into different sorts of systems, depending on the type of phenomenon they are investigating (Elliott-Graves, 2020b). So, for example, ecological systems can be populations, meta-populations, communities, ecosystems and so on. Parts of populations typically are individuals, parts of ecosystems can be populations or species, abiotic factors (e.g. nitrogen) and so on. Notice that for each system the parts need not occur at the same hierarchical level but can cut across levels (so one part of an ecosystem can be a population, another part can be an element and another can be a species). The second characteristic is interaction: the multiple parts of the system interact with each other. Here, Levins mentions demographic interactions, which occur within and between populations, but given his inclusion of environments as a characteristic, I think we can safely assume that he would also count interactions between individuals (or populations) with parts of the environment as relevant. The third characteristic is heterogeneity, which Levins mentions in relation to species (i.e. many different species), individuals within a population (genetic, physiological and age diversity) and types of environment.

The next question to ask is what is the relationship between these three characteristics? This is where things really start getting interesting. On the face of it, it seems that for Levins, all these characteristics are aspects of complexity. So, all these three characteristics together make up the notion of complexity. All these aspects *together* cause the trade-offs. However, John Matthewson argues that complexity and heterogeneity are distinct concepts. In his 2011 paper, which critiques but also builds on Levins's trade-off account, Matthewson points out that 'complexity' can be defined in many different ways, including the 'technical definition': 'it has many interacting parts and perhaps exhibits some kind of emergent behaviour' (2011, p. 331). He then seems to adopt this technical definition, which centres on the 'many interacting parts' characteristic (2011). Moreover, he argues that complexity alone cannot account for the trade-offs that biologists routinely face. After all, scientists in other fields that study complex systems, such as physics and chemistry, do not seem to face such extensive trade-offs (see also Justus 2005, 2021, section 2). Matthewson asks us to consider a group of Airbus A330s. They are clearly complex systems, in the sense of being made up of many interacting parts. Yet we can construct successful models that simultaneously capture all the relevant variables, apply to all Airbus A330s, and provide precise predictions about their flight trajectories. In other words, they are maximally general, realistic and precise. There must be some characteristic other than complexity, that causes the trade-offs between desiderata in biological systems.

This is where heterogeneity comes in. Biological systems (e.g. ecosystems), like aeroplanes, are made up of many interacting parts, but unlike aeroplanes, each ecosystem is unique. A marine ecosystem and a forest ecosystem might have similar trophic levels, but the species in each level are very different (in the aeroplane analogy, this would be akin to a situation where the engines of various A330s were actually different). Thus, the knowledge we gain from examining one ecosystem does not transfer to others, even if they seem similar. For example, a model of the marine ecosystem might not generalise to a forest ecosystem and vice versa. In short, heterogeneity magnifies the trade-offs between desiderata, by restricting a model's ability to generalise (a similar view of heterogeneity as magnifying certain trade-offs can be found in (Weisberg, 2004)).[9]

I propose two refinements to Matthewson's account of heterogeneity. The first is a shift in focus from *ontological* to *causal* heterogeneity. Matthewson's conception of heterogeneity is ontological, in the sense that it refers to differences in the nature of systems' parts. Thus, Airbus A330 parts (e.g. the engine, the wings) are the same type of thing across different individual Airbus A330s,

---

[9] I should note that these are not the only options regarding the concepts of complexity and heterogeneity. Another view is one where complexity (understood as multiple interacting parts) is not merely distinct from but antithetical to heterogeneity. The claim is that despite the diversity within the parts of a complex system, there must be sufficient homogeneity between the parts, so that they obey the same physical laws and so that complex behaviours can emerge. This view is proposed by Ladyman et al. (2013) and was clearly conceived in a non-biological context, though its proponents believe it also applies to biological and social systems. Just as in physical systems, particles must be 'comparable in size and weight' so that complex behaviours can emerge, so in biology, cells 'before they form multi-cellular organisms are indistinguishable' and in social systems, social structures 'have to be similar in character, behavior or rules obeyed' (p. 57). From an ecological perspective, this view is rather odd. A strong reading of this view is clearly false, as it is incompatible with biological reality. On such a reading, ecosystems could not exist: as the parts of an ecosystem are highly diverse (multitudes of different species and abiotic factors) their component parts should not be able to interact. Thus, instead of being 'prototypical' examples of complex systems, ecosystems would be reduced to figments of ecologists' imaginations. A more charitable interpretation of the view could be that Ladyman et al. are not denying that diversity of parts exists, but that it is not as important as the underlying similarities between parts of systems.

whereas the components of ecosystems (e.g. plants and animals) are different types of things: each ecosystem has a unique collection of species. The problem is that ontological differences between systems are neither necessary nor sufficient for heterogeneity. For example, the phenomenon of niche overlap occurs when different species such as sardines and anchovies fulfil very similar functions within an ecosystem, so the populations can be substituted without affecting the functioning of the marine ecosystem (Ricklefs and Miller 2000). Here, ontological differences are not sufficient for heterogeneity, as ontologically diverse species are causally *homo*geneous.

Even more interesting, for our purposes, are the cases where ontological heterogeneity is not necessary for causal heterogeneity, that is, when ontologically similar or even identical species are causally heterogeneous. Consider the case of plant-soil feedback (PSF) (Figure 2). Plants interact with microbes in the soil, for example, arbuscular mycorrhizal fungi and nitrogen fixing bacteria. These interactions can be beneficial to the plant's growth (positive feedback), for example, displacing plant pathogens or detrimental (negative feedback), for example, less access to nitrogen (Figure 2A). PSF can affect the outcome of competition between species, as positive feedback can confer a competitive advantage to a plant, while negative feedback can significantly reduce a plant's ability to compete (Figure 2B). The interesting thing about PSF is that the type of feedback (positive, neutral or negative) can vary even when these systems are made up of the same species of plants and soil microbes (Klironomos, 2002; van der Putten et al., 2013). For example, in field experiments, the sign of the PSF has been known to reverse in a single community, changing from positive to negative feedback (Casper and Castelli 2007; Klironomos 2002). That is, the same soil biota that were boosting plant growth begin to hinder the growth of those same plants. The opposite switch has been observed in plant invasions, where plant species experience negative feedback with soil microbes when they first move to a new area, but then begin to experience positive feedback in the same system (van der Putten et al., 2013).

In short, it is the *behaviour* of systems and their parts that is relevant, rather than the *makeup* or *nature* of these systems. Of course, there will be many cases where the ontological differences between systems or their parts is the reason why they behave differently. In other words, ontological heterogeneity often contributes to or even entails causal heterogeneity. Still, even in these cases, it is the causal heterogeneity that is most relevant for studying the behaviour of ecological systems. Moreover, causal heterogeneity is the more useful of the two concepts: it encompasses all the instances where ontological heterogeneity leads to causal heterogeneity but excludes all those where ontological heterogeneity leads to causal *homo*geneity.

Figure 2 Plant-soil feedback. (A) Plants interact with microbes in the soil, leading to positive, neutral or negative feedback; (B) PSF can affect the outcome of competition between species, as positive feedback can confer a competitive advantage to a plant, while negative feedback can significantly reduce a plant's ability to compete

The second refinement concerns the units of heterogeneity. Matthewson focuses on 'inter-system heterogeneity', that is, heterogeneity between different systems. Each Airbus A330 or ecosystem is a complex system, and heterogeneity occurs between these systems. Thus, two Airbus A330s are homogeneous with respect to each other whereas two ecosystems are heterogeneous. However, causal heterogeneity can also manifest *within* each system, as differences between the parts of a system can have different effects. I call this *intra*-system heterogeneity (Elliott-Graves, 2018). Both types of heterogeneity are important because each causes a different type of problem for ecological research. Inter-system heterogeneity hinders generalisation across different systems and intra-system heterogeneity hinders generalisation within a system. This usually occurs when the future behaviour of a system is different to its past behaviour, which predominantly affects our ability to make predictions. I will examine the effects of inter- and intra-system heterogeneity in the next section (3.1 and 3.2).

So much for multiple parts, interaction and heterogeneity. What about the other characteristics? I believe that they are not as important *for this context*, that is, identifying the epistemic effects of ecological complexity, such as trade-offs. Let me clarify. The trio of multiple parts, interaction and (causal) heterogeneity can capture the main aspects of the other characteristics, *in this context*. In other words, the other characteristics can be seen as special cases of one, two or all three of the trio. For example, aspects of non-linearity, such as sensitivity to initial conditions and feedback can be seen as instances or aspects of causal heterogeneity. Changes in initial conditions cause systems to behave differently, while feedback does the same because of magnification of an effect. Both are captured by causal heterogeneity (see discussion on PSF above and example 2 – Emerald Ash Borer). I should note that my claim is not that the other characteristics are superfluous or irrelevant *generally*, merely that we do not need to discuss them separately in the context of their epistemic effects – we can account for these effects by focusing on the trio.

2.5 Ecological Complexity

To recap the main points of the discussion so far, Levins believed that complexity should be understood as a combination of three characteristics, multiple parts, interaction between the parts and heterogeneity of the parts themselves. Matthewson argued that despite Levins's inclusion of heterogeneity in the notion of complexity, what people usually mean when they use the term 'complexity' is multiple interacting parts. This is a problem, because leaving heterogeneity out of the picture means that our notion of complexity does not account for the differences between biology and other sciences in terms of the extent and magnitude of trade-offs. However, Matthewson goes a bit too far in the other direction, as he claims that heterogeneity alone explains the trade-offs. My view is that heterogeneity generates new trade-offs and magnifies existing ones, so it should be viewed as an important characteristic of complexity, but not as a concept entirely independent of complexity. I further refined Matthewson's notion, arguing that causal rather than ontological heterogeneity accounts for the magnification of trade-offs.

Thus, finally, we have our answer: ecological systems are complex in the sense that they are made up of *multiple causally heterogeneous interacting parts*. I should quickly point out, once again, that I am restricting my analysis to this particular context: understanding the epistemic difficulties in ecological research. It is *within this context* that this definition is meant to apply. There may well be other contexts in which other notions of ecological complexity turn out to be more useful. But if we are trying to understand why ecological systems are difficult to investigate, it is because they are made up of many causally heterogeneous interacting parts.

3 What Are the Effects of Ecological Complexity?

Now that we have a characterisation of ecological complexity, we can move on to the next issue at hand, namely what difficulties ecological complexity creates for ecological research and how or why it creates them. The discussion in the previous two sections implied that the primary difficulty is for the creation of *generalisations*. Here, I will examine this difficulty in more detail and also explain why generalisations are sometimes possible (Section 3.1). I will then examine how the failure to generalise affects ecologists' ability to make precise and accurate *predictions* (3.2). In Section 3.3, I will show that these problems are not merely theoretical but flow into the domain of applied ecology, as they decrease the likelihood of successful *interventions*.

3.1 Generalisation

3.1.1 Why Generalise?

Why do scientists and philosophers value generalisations? Generalisation is a key dimension of scientific theorising, for scientists as well as philosophers. For example, Richard Feynman (1995, in Mitchell 2000, p. 245) stated that 'science is only useful if it tells you about some experiment that has not been done; it is no good if it only tells you what just went on. It is necessary to extend the ideas beyond where they have been tested'. In other words, generalisations can help scientists gain insights that go beyond their current investigation. More specifically, generalisations are valuable in three related ways: (i) they reveal information about the world (ii) they underwrite explanations and (iii) they form the basis of predictions.[10]

Generalisations reveal information about the world because of the process by which they are formed, that is, *abstraction* (Cartwright, 1989; Elliott-Graves, 2020b; Jones, 2005; Weisberg, 2013).[11] The process usually starts by identifying which factors are idiosyncratic to particular systems and which are common to many systems. The idiosyncratic factors are then omitted and what remains, constitutes the generalisation (Cartwright, 1989). The idiosyncratic factors of a system are thought to be 'mere details', or noise, in the sense that they do not contribute significantly to the dynamics or behaviour of the system. In contrast, the common factors are thought to be the 'real causal factors' of the system that actually give rise to its behaviour and dynamics. Thus, successful generalisations provide scientists with knowledge about how the world really is (Strevens, 2004).

This allows scientists to 'virtually reduce' the complexity of a system, which makes the system easier to study. If we go back to Levins's framework (Figure 1), this is a motivation for adopting type I models, which sacrifice realism for the sake of generality and precision. Omitting the noisy details from a system means that there are fewer factors that scientists need to take into account, hence they can study the systems *as though* they have fewer parts and properties. Thus, there are fewer variables and parameters to measure or calculate and fewer dynamics to identify. Moreover, there are fewer possibilities of introducing errors to the models, as is the case with type II models.

These ideas were incorporated in many influential accounts of scientific explanation: to provide a scientific explanation used to be synonymous with demonstrating how a particular phenomenon is an instance of a more general pattern, while *scientific* theories were those that subsumed many disparate phenomena under one framework (Hempel & Oppenheim, 1948; Kitcher, 1981, see also discussion in Douglas, 2009 and Strevens, 2004). Here, the simpler the theoretical framework and the greater the number of phenomena it

---

[10] All three attributes are present in most philosophical accounts of laws and explanations, though philosophers differ with respect to how closely they believe these three attributes are connected and whether or not one takes precedence. For Hempel & Oppenheim (1948), for example, these three attributes are all sides of the same coin: laws are generalisations that are true and not accidental. Explanations and predictions are directly deduced from these laws. On causal accounts of explanation, generalisations are valuable because they form the basis of explanations, and it is through explanations that we gain knowledge about the world. Little is written about predictions in these accounts, though it is usually understood or implied that generalisations also form the basis of predictions (Douglas, 2009).

[11] On some views, e.g. Levy (2018), abstraction and generalisation are the same. I believe that they are distinct. Abstraction is a necessary prerequisite for generalisation, but abstraction does not necessitate generalisation (Elliott-Graves, 2020b; 2022).

could explain, the more explanatorily powerful it was. More recently, for example, in causal accounts of explanation such as Woodward's, generalising does not constitute an explanation, but generalisations form the basis for explanations, as they describe causal regularities between variables. Here, to explain an event or phenomenon is to show that it is caused by another event or phenomenon and that this relationship is invariant or stable within certain parameters (see also Section 3.1.2).

Finally, generalisations are valuable because they form the basis for predictions. I will turn to prediction in Section 3.2, after discussing how ecological complexity affects generalisations.

3.1.2 Complexity, Laws & Generalisation

In order to understand how complexity affects generalisation, it is useful to examine the debate about laws in (evolutionary) biology. On the traditional philosophical account, laws of nature have three main characteristics: (i) universality, that is, they cover all of space and time, (ii) truth, that is, they are exceptionless and (iii) natural necessity, that is, they are not accidental (Mitchell 2003, p. 130).[12] The first and second characteristics concern the scope of generalisations. Universal generalisations exist throughout all space and time. They are true (i.e. exceptionless) if they hold in all instances within their defined scope. Thus, even if a generalisation is not universal, it can still be exceptionless within a specified domain. For example, it is likely that in the first few minutes after the Big Bang, only helium, deuterium and lithium were formed. All other elements were formed later, as stars developed (Mitchell, 2003 p. 137). Thus, a law stating that no uranium-235 sphere is more massive than 55 kg is only applicable to the time where uranium-235 actually exists.

The third characteristic aims to distinguish between generalisations that are necessary and those that are merely accidental. A favourite example of an accidental generalisation for philosophers is 'all the coins in Goodman's pocket are made of copper' (Mitchell 2003, p. 136). There is nothing about Goodman's pocket or the nature of coins or copper that affects whether or not the statement is, in fact, true. If it turns out to be true, then it is true accidentally. This has important pragmatic effects for scientific practice. A scientist could not explain why all the coins in Goodman's pocket are copper, nor could she predict whether a new coin placed in Goodman's pocket

---

[12] Mitchell also identifies *logical contingency* as a characteristic of laws. This means that laws have empirical content, i.e. they are not merely definitional. There is general consensus in philosophy of science that laws in biology satisfy this criterion, so I will not discuss it any further (but see Sober (2011).

would also be copper. In contrast, the statement in the previous paragraph about the maximum mass of uranium is not accidental: a uranium sphere above a certain mass would cease to be stable and would therefore cease to be a uranium sphere. This also constitutes an explanation for the law and could yield predictions about uranium spheres.

Most candidate laws in biology fail on all three counts (Mitchell 2003). The typical example of a biological law is Mendel's 'segregation law of gametes'. It states that for each pair of an organism's genes, 50 per cent of an organism's gametes will carry one representative of that pair, and 50 per cent will carry the other representative of the pair (Mitchell 2003, p. 139). This law is far from universal. It only applies to sexually reproducing organisms, which are a relatively small subset of or all organisms that exist in a small corner of the universe for a small amount of time, in the grand scheme of things. Yet even within these parameters, it is not exceptionless. In cases of meiotic drive, one gene doubles or triples its gamete representation so that it has a higher likelihood than its pair to represented in the offspring (Mitchell 2003, p. 149). Thus, Mendel's law is contingent on the absence of meiotic drive. Finally, Mendel's law is also contingent in a different sense. All evolutionary biology is contingent on its own history. That is, if we 'rewound the history of life' and 'played the tape again', species, phenotypes and body plans would be quite different (Gould 1989, in Mitchell 2003, p. 148). There is nothing necessary about the way the biotic world has evolved. In Mitchell's words 'The causal structures that occupy the domain of biology are historical accidents, not necessary truths' (2003, p. 148).

Rather than concede that biology has no or very few laws, some philosophers proposed that we change our conception of law to accommodate biological complexity (Mitchell 2003; Woodward 2001, 2010). They argued that these dichotomies (necessary/accidental, universal/contingent) are not useful. The point of generalisations is that they help scientists achieve various goals, such as explaining and predicting natural phenomena or intervening on the world. Thus, the relevant characteristic for generalisations is *stability* (Mitchell, 2003)[13] or *invariance* (Woodward, 2001, 2010). Stability is a measure of the range of conditions that are required for the generalisation to hold. Invariance is a subset of stability that focuses on direct causes[14]: 'a generalization is invariant

---

[13] On Mitchell's account, stability is one of three dimensions of scientific law, the others being strength and abstraction (Mitchell 2003, p. 146). I focus on stability, as this is the dimension where ecological laws fall short, due to causal heterogeneity.

[14] There is some disagreement between Woodward and Mitchell concerning the extent to which these concepts are similar (Mitchell 2003, Woodward 2001, Raerinne 2011). The differences between the two views are not relevant here. Causal heterogeneity in ecology renders generalisations unstable and invariable.

if and only if it would continue to hold under some range of physical changes involving interventions' (Woodward 2001 p. 4). For Woodward, an intervention is a hypothetical manipulation of a variable in an ideal experimental setup, which would allow us to determine whether that variable caused another (for example, turning a knob on the radio causes the volume to increase or decrease).

An example of a generalisation that is invariant/stable, but only contingently so, is Hooke's law (H) when applied to a spring (Woodward 2001, pp. 10–11). The generalisation (H) is expressed as $F = -kX$, where X is the extension of the spring, F the restoring force it exerts and k a constant characteristic of springs of sort S. H correctly describes what the restoring force of the spring would be under a series of experimental manipulations of the extension of the spring, but only within a certain interval. Put simply, if we pull too hard, the spring might break and the generalisation H will break down. Even though H is not invariant under *all* interventions, it is still invariant under *some range* of interventions. Thus, the generalisation is useful, even though it is not universally invariant: it explains why the spring returns to its original position when it does, and also why it does not (the force was too great).

Ecology had its own parallel debate about laws, which mirrored the developments of the complexity/laws debate in biology. Early ecologists were at pains to uncover general laws akin to those of physics, so as to convince the scientific community that ecology was a true scientific discipline, even adapting simple, general models from physics for this purpose (Kingsland, 1995). Various candidates for lawhood have been proposed, including exponential/logistic growth, allometric metabolic/body weight relationships and species/area relationships (Colyvan & Ginzburg, 2003; Lange, 2005; Turchin, 2001). However, none of these potential laws have been proven to be universal or exceptionless (Beck, 1997; Shrader-Frechette & McCoy 1993). Just as in the case of evolutionary biology, the absence of such laws was taken by some as an indication that ecology is not a truly scientific discipline (Lawton, 1999; Peters, 1991). Just like evolutionary biology, however, others believed that the problem lay in the conceptions of laws. Thus, the conception of laws was weakened to account for complexity (Colyvan & Ginzburg, 2003; Cooper, 1998; Lange, 2005; Linquist et al. 2016).

For example, Linquist et al. (2016) have formulated an account inspired by Mitchell's and Woodward's accounts of invariance and stability. They identify three dimensions of *resilience*. A generalisation is *taxonomically* resilient if it is stable across a different number of species or higher-level taxa. *Habitat* resilience occurs when a generalisation is invariant across a broad set of regions or biological contexts, while *spatial* resilience occurs when a generalisation remains invariant at the scale of whole organisms, molecular systems, and genomic communities. They use meta-analyses to investigate the extent to which various generalisations are resilient across these three dimensions. For example, they found that the generalisation 'Habitat fragmentation negatively impacts pollination and reproduction in plants' is resilient across 'five distinct habitats and across 89 species from 49 families' (p. 129).[15] They believe that these types of generalisations qualify as laws.

The aim of this discussion in this section was to reiterate and highlight an issue that has already been established by philosophers of science, that is, that complexity affects generalisability – in the sense that generalisations in biology, ecology included, do not have the same scope as generalisations in other disciplines. I also broadly agree with those philosophers who argue for a revised notion of laws that better reflects biological complexity. There may even be cases in ecology where generalisations could qualify as laws in the revised sense. However, I also believe that ecology is rife with cases where generalisations do not fulfil even those revised requirements, that is, where they are not invariant, stable or resilient. What happens in those cases? Here I will move away from the debate about laws, as I am not interested in whether we should adjust our notion of laws even further so that these generalisations could qualify, but in *why* they do not satisfy the revised notion of laws in biology. I address these issues in the remainder of Section 3.

3.1.3 Causal Heterogeneity and Generalisations

Generalising involves comparing and contrasting various systems and distinguishing between the common factors and the idiosyncrasies. Ordinarily, generalisations are possible because the factors that are common between systems are also the factors that are causally relevant. The idiosyncratic factors are irrelevant – mere noise. General models or theories can thus include only those factors that are relevant and omit all the idiosyncratic factors. The problem is that in causally heterogeneous systems, idiosyncratic factors are not mere noise, but causally relevant for the functioning of the system. This is what creates difficulties for making generalisations. The factors that give rise to a phenomenon in one system might not be the same as those in another system, even if the two systems seem very similar (e.g. have similar sized populations of

---

[15] I should note that not everyone agrees. Leonore Fahrig (Fahrig, 2017) argues that the data only shows that habitat *loss* affects resident populations (and hence biodiversity), whereas habitat *fragmentation* without loss often does not have the same negative effects.

the same species). Thus, the knowledge we gain from studying one system is not always adequate for understanding the behaviour of other systems, or even of the same system at a different time.

But is this not already covered by the traditional notion of complexity? In the previous section, we saw how recognising that biological systems are complex led to a revision of our conceptions of laws, to generalisations that are contingent and allow for exceptions. I agree with Mitchell, Woodward and so on, but I also believe they do not go far enough. Invariant generalisations are common when systems are merely complex, whereas causal heterogeneity makes generalisations unstable. Recall Matthewson's example of the aeroplanes: each aeroplane is a complex system, yet the factors that affect how one Airbus A330 gets into the air are the same as those that affect how the other Airbus A330s get into the air. Differences between them, for example, the logos of each airline are irrelevant for the purpose of getting into the air and can thus be safely ignored.

When heterogeneity is added to the mix, even this level of invariance ceases to hold. To illustrate, I will adapt Woodward's spring example to an ecological context. Insects (like most organisms) have a certain range of temperatures that they can tolerate. In cold climates, the lower end of the threshold is especially important because it affects insects' overwintering strategy (e.g. freeze tolerance or freeze avoidance) (Sinclair & Jako Klok, 2003; Sinclair & Vernon, 2003). This, in turn, can be used to predict an insect population's abundance and spread, which can form the basis of interventions
(e.g. if the insect in question is a pest). The lower lethal temperature (LLT) for an insect can be measured in the laboratory.

Such an experiment could yield the generalisation (T): W can tolerate temperatures between -30 and +40 C°, where W is a spatially defined population of a particular species of insect. In Woodward's terminology, generalisation T is invariant within the domain −30 to +40 C°. However, it turns out that the very act of lowering the temperature (i.e. intervening on the system) can change the range of temperature that the insect can tolerate (Kaunisto et al., 2016; Marshall & Sinclair, 2012; Sinclair et al., 2003). For example, if the temperature is lowered to −30 C° at time t_1 the insect can survive, but then it will only survive down to −25 C° at another intervention, at time t_2. In fact, it is possible that repeated exposures to low temperatures change the LLT. For example, at the interventions at time t_1, t_2, t_3 and t_4 (where the temperature reaches −30 C°) there is no change in the LLT, but there is at t_5. To make matters worse, there is no set number of exposures after which Ws lose their ability to tolerate −30 C°. It could be after the fifth intervention, but it could also be after the third, the sixth, the ninth, and so on.

How should we think of generalisation *T*? At first glance it seems very similar to Hooke's law, as it holds within a certain range. We can intervene on the system and determine the range at which the generalisation is invariant. However, at some point the pattern ceases to exist. This is not accidental, random or even external to the system, but because of the very interventions that are used to establish the range of invariance. It is not equivalent to a fire starting in the laboratory and melting the spring so that there is nothing that the restoring force could act on. It is the equivalent of a non-defective spring suddenly breaking after being pulled a few times with a force well within the range that it can withstand.

One option is to say that *T* is not, in fact, a generalisation, as it turns out that it is not invariant across the range −30 to +40 C°. But this seems a bit odd. The generalisation *was* invariant for a while, then it stopped being invariant. It is only after a number of repeated exposures to −30 C° that *W*s cease to tolerate that temperature. In fact, it is because they *can* survive to −30 C°, that they cannot do so in the future. Another option would be to constrain the range of *T* to −25 to +40 C°. Then, the generalisation would be invariant even after repeated exposures to −30 C°. But this is also rather odd. *W*s regularly survive to −30 C°. An explanation or prediction based on the constrained generalisation would probably be inaccurate. Consider, for example a situation where *W*s are pests and we wanted to know if a particular agricultural crop would be at risk from *W*s. A farmer will use a certain pesticide after the winter thaw, but only if the temperature does not drop below a certain point. If we constrain the range of T to −25 to +40 C°, then the farmer will not spray if the lowest winter temperature was −26 C°. But this will result in a catastrophe for the crops, because *W*s *can* survive to −26 C°, if that is the lowest winter temperature.

The point of this example is to show that generalisations in ecology are special when compared to generalisations in other parts of biology. They are special in the sense that they are *extremely* contingent. Contingent generalisations in other parts of biology may nevertheless display stability or invariance, whereas the contingency in many ecological generalisations is so extensive that it precludes even this. There are three points to reiterate and clarify. First, complexity in the sense of having multiple parts cannot account for the extreme levels of contingency found in many ecological systems, whereas the revised notion, which includes causal heterogeneity can account for it. Second, the difference in contingency is a matter of degree not a difference in kind. That is, generalisations in ecology are just *much more* contingent than they are elsewhere. Nevertheless, this extra contingency has far-reaching epistemic implications, because it affects our ability to generalise, and consequently to predict and intervene on ecological systems. Third, it is not the case that generalisations are completely absent, or merely instances of ecologists being *mistaken* in their identification of patterns. Of course, there will instances of mistakes in ecology, just as in other sciences. But these are not such cases. These are cases where there *is* a pattern, and ecologists identify it correctly. The problem is that the pattern only holds in a very small range of space and/or time. This is what I mean by the claim that patterns are *ephemeral*.

When faced with this level of contingency and ephemeral patterns, scientists can (i) restrict the scope of the generalisation, (ii) bite the bullet and retain the generalisation even though it only applies well to some systems or (iii) weaken the generalisation to accommodate the peculiarities of each system. Ecologists are increasingly beginning to favour option (i), which I will discuss in Section 4. Historically, however, ecologists have opted for options (ii) and (iii), because of the importance placed on generalisation in ecology and the associated worries that a discipline without generalisations is not truly scientific (Section 1.2). The problem with option (ii) is that it is strictly speaking false: the generalisation only holds within a subset of the cases that it purports to. The problem with option (iii) is that it reduces the value and usefulness of the generalisation because it renders it incapable of yielding explanations. I will illustrate the pitfalls of adopting these two options, by outlining the rise and fall of the keystone species theory.

- *Example 1. Keystone Species:*
  The *keystone species* concept was coined by Robert Paine in 1969, to describe the importance of some species for the overall functioning of an ecological community. It was based on two experiments of intertidal communities (one on the Pacific Coast of Washington and one on the Great Barrier reef), where the removal of predators led to local extinctions of other species in the community and the deterioration of the community's structure (Cottee-Jones & Whittaker, 2012). The important innovation was that species higher up on the food web, with comparatively little biomass when compared to those on lower trophic levels, determined and maintained the structure and composition of the community. The concept quickly became entrenched in the literature and was used to explain a number of events and phenomena in ecological communities (Beck, 1997; Mills & Doak, 1993). It has since become a standard entry in textbooks (Cottee-Jones & Whittaker, 2012). This, together with its use in framing and gathering support for conservation

policy, increased its popularity outside academia, ascribing it the dubious honour of 'buzzword' status (Barua, 2011; Valls et al., 2015).

However, subsequent investigations cast doubts as to the generality of the keystone species concept. Experiments have shown that various 'keystone' species demonstrate wildly different behaviours. In quite a few cases, their presumed effects were much smaller or even entirely absent from many systems (Beck, 1997). Incidentally, keystone effects were also absent in communities similar to those that Paine had studied (i.e. intertidal communities in Oregon containing the starfish *Pisaster ochraceus*, the same predator removed in Paine's original experiment) (Cottee-Jones & Whittaker, 2012).

The predominant response by many ecologists was option (iii), that is, to adapt the concept and make it more inclusive, so that it could apply to many diverse systems and accommodate the various peculiarities of potential 'keystone' species. Thus, a keystone species became any species that had a disproportionate effect on the community given their biomass, irrespective of their trophic level. Other definitions included any species that has a disproportionate effect given their abundance, any species that has a strong influence on the community. In the 1980s, the concept was expanded even further to include plant species that had functioned as prey or mutualists (Cottee-Jones & Whittaker, 2012).

Relaxing the stringency of the definition allowed scientists to encompass more particular cases under the keystone concept, but at a price. The concept of keystone became increasingly watered down. As Mills et al. (1993) pointed out, the concept was applied to so many diverse species, that it was no longer useful for explaining community interactions. If all these diverse behaviours of species within their communities could count as 'keystone' behaviours, then what does the keystone concept actually pick out? What is special about a keystone species? Thus, merely classifying a species as a keystone provided no real information about how it will affect a community. This was especially problematic because it affected the efficacy of conservation efforts (Cottee-Jones & Whittaker, 2012; Mills & Doak, 1993). In addition, the watering down of the concept led to the concept becoming popular outside academia, which, in turn, led to even more broad and vague uses of the term (Barua, 2011).

This led to a backlash within the scientific community. One suggestion was to adopt a single but broad definition (e.g. 'a species that has demonstrable influence on ecosystem function'), yet this also suffered from the same lack of explanatory power (Barua, 2011). Some ecologists have decided to accept this and now view the keystone species concept as a metaphor or heuristic, whose value is mainly pedagogical (Barua, 2011; Valls et al., 2015). The remainder seem to have seriously embraced option (ii), that is, restricting the definition of a keystone species in the hope of salvaging its explanatory power (Cottee-Jones & Whittaker, 2012). The idea is to go back to the drawing board and start from the beginning, using experimental and comparative methods to determine if there is a viable definition of keystone species, and which (admittedly fewer) species truly exemplify the relevant keystone characteristics (Cottee-Jones & Whittaker, 2012).

This example illustrates a theme that is far from unique in ecology. A phenomenon is observed in one or more systems, experiments and models are carried out and the results seem promising. A theory is constructed, which identifies the general, underlying mechanisms that give rise to the phenomenon. The scope of the theory is expanded to encompass a number of similar phenomena and/or phenomena that are consequences of the original phenomenon. Then the counterexamples start cropping up and gaining in numbers, so the theory is either tweaked to the levels of being trivial or slowly fades away (Beck, 1997).

Of course, causal heterogeneity does not preclude *all* generalisations. If there really were no generalisations in ecology, scientific ecological knowledge would not involve any theories, just long lists of particular observations, while experiments would identify unique causal relations that disappeared after the conclusion of the experiment. The question is, what generalisations are possible and why do they persist despite extensive causal heterogeneity?

3.1.4 When and Why Do Some Ecological Generalisations Hold?

There are two types of cases where generalisations in ecology hold (see also Raerinne 2011). The first is when the generalisations are exceedingly *modest*. Scientists are sometimes able to identify common causal factors across a few systems or for a limited amount of time. When scientists apply their generalisations within these strict limits, their generalisations hold. I will examine examples of such generalisations in Section 4.2.

The second type of case where ecological generalisations hold comes down to sheer luck.[16] It just so happens that among these complex heterogeneous

---

[16] I should note that when I refer to *luck*, I do not mean that the generalisation is *accidental*. Recall that being accidental is a characteristic of non-law-like generalisations, such as 'all the coins in Goodman's pocket are made of copper', as opposed to 'no uranium-235 sphere is larger than

systems, sometimes the relevant causal factors for a particular system, state or behaviour, *are just those factors that are common across systems*. Delving deeper into this point is helpful because it elucidates the effect of causal heterogeneity on generalisation. Ecological systems are governed by many causal factors. Some of these are common across systems. A good example of this is density dependence, the effect that the size of a population has on its members (see also Box 1A and Section 4.1). All ecological populations live in finite geographical regions, with finite resources. At some point in their growth, growing further will be limited by the diminishing availability of these resources. This is a factor that affects all ecological populations. However, it is not always *relevant* for the particular phenomenon under investigation. In some cases, this may be because the population has not yet reached the size where resources are starting to dwindle. The causal mechanism is there, but it has not yet come into effect. More often, a factor becomes irrelevant because its effect is reversed or overshadowed by another more powerful causal factor. Thus, for example, habitat loss or poaching might be the relevant causal factor that explains the size of a particular population A. It is not that density effects have ceased to operate, but that they are overshadowed by a much stronger effect, *in this particular context* (i.e. explaining the size of population A at time t_1).

How does this relate to generalisations? Sometimes, scientists are lucky in that the dominant causal factor is common to two or more systems. Unfortunately, this happens most frequently in cases of human action (such as poaching or habitat loss), whose strength trumps the effect of most non-human causal factors. Thus, scientists can expect that effects of habitat loss will generalise to many diverse systems (across different ecosystems, communities, species, etc.). Somewhat less frequently, there are common causal factors across systems because of the absence of stronger heterogeneous causal factors. Thus, for example, density dependence effects are strong for rare tropical trees (Comita et al., 2010), voles and lemmings (Stenseth, 1999), juvenile survival of large herbivores (Bonenfant et al., 2009) and magpies (when food hoarding) (Clarkson et al., 1986) (see also Section 4.1).

To sum up, causal heterogeneity explains the difficulties of generalising, and its occasional absence explains why generalisations are sometimes possible.

---

55 kg'. When I claim that scientists are lucky, I mean that the generalisations are invariant and continue to hold. In this sense, they are more law-like than accidental. Nonetheless, I believe that in causally heterogeneous systems, it is rare, or at least not commonplace for generalisations to be invariant in this way. As this makes the scientists' job easier, scientists who do find invariant generalisations can legitimately be termed 'lucky'.

But the problems do not stop here. Failed generalisations lead to difficulties in making predictions, the subject of the next section.

3.2 Prediction

3.2.1 The Traditional Account of Scientific Prediction

Scientific predictions are statements about events, phenomena or behaviours of a system, whose truth value is not known at the time they are made. Predictions are useful for two reasons, because they can help us test and confirm scientific theories, and because they guide interventions on phenomena in the world (Barrett & Stanford, 2006; Lipton, 2005). Philosophers of science have predominantly focused on *confirmatory* predictions, because of their role in testing and comparing scientific theories (Barnes, 2018).[17] An example often used in textbooks is Mendeleev's predictions of new elements. When he classified the elements into the periodic table according to atomic weight, there were some gaps. Mendeleev predicted that these gaps would be filled by hitherto unknown elements, with specific physical and chemical properties (e.g. atomic weights, acidity, specific gravity). Eventually the elements scandium, gallium and germanium were discovered, and their physical and chemical properties matched Mendeleev's predictions. This discovery is said to have 'vindicated' Mendeleev, in the sense that it showed that his classification of the elements was better than various alternatives (Scerri, 2006). Thus, the new elements were said to confirm Mendeleev's background theory.[18]

In order to be successful, confirmatory predictions must possess certain characteristics. The most obvious of these is *accuracy,* that is, the predictions should turn out to be true. However, accuracy alone is not sufficient for theory confirmation, as accurate predictions are sometimes too easy to come by. A prediction could come out true because of pure chance; for example Mendeleev could have just guessed that new elements would be discovered without reference to the periodic table. Alternatively, a prediction could be obviously true, for example if Mendeleev predicted that there may or may not be other elements. Finally, accurate predictions can come about due to more nefarious reasons, such as predicting an event that is already known to be

---

[17] I will focus on confirmatory predictions in this section and address the other role of prediction (which I call *applied* prediction) in Section 3.3.

[18] I should note that this potted history of the periodic table is greatly oversimplified. In reality, the acceptance of Mendeleev's theory was much more complicated; scholars have even questioned the role of the predictions for its acceptance (Scerri & Worrall, 2001). Still, the 'standard story' is worth mentioning even though it may be strictly speaking false, because it shows just how deeply entrenched the traditional notion of prediction is in the literature.

true, for example, if Mendeleev had already discovered the new elements before predicting their existence.

Therefore, a confirmatory prediction should also be *risky*, that is, it should be possible that it turns out to be false. There are two ways of increasing the riskiness of a prediction. The first is to ensure that it is *novel*. Novel predictions are those that are made without knowledge of the facts being predicted (Douglas & Magnus, 2013; Hitchcock & Sober, 2004).[19] The idea is that novel predictions cannot be faked or tweaked to fit existing data, hence they provide the best way to test and confirm scientific theories. Not everyone agrees with this view (which is called *strong predictivism* in the literature). On the other side of the debate are those who believe that *accommodating* existing data provides sufficient empirical support for scientific theories (Hitchcock & Sober, 2004).[20] A third view is *weak predictivism*, which states that novel predictions are valuable because they correlate with other epistemic virtues such as explanatory power or simplicity (Hitchcock & Sober, 2004).

The second way to increase the riskiness of a prediction is to increase its *precision*. Precision means how *finely specified* a prediction is. For example, the prediction 'the temperature at noon tomorrow will be high' is less precise than 'the temperature tomorrow at noon will be between 25 and 35°C', which is less precise than 'the temperature tomorrow at noon will be 31°C'. Precision trades off with accuracy: the more precise a prediction is, the likelier it is to be inaccurate. This explains how precision increases riskiness: the smaller the range of values predicted, the larger the range of actual values that will render the prediction inaccurate and vice versa. For example, if the temperature at noon tomorrow turns out to be 26°C, the first two predictions ('temperature will be high' and 'temperature will be between 25 and 35°C') are accurate, because the ranges they specify include the actual temperature, whereas the most precise prediction is inaccurate. If the actual temperature turned out to be 24°C, only the least precise prediction would be accurate.

To sum up this section, the traditional account of scientific predictions views predictions as means to test and confirm scientific theories. These confirmatory predictions are valued in terms of accuracy and riskiness, which are achieved through novelty and precision. As we shall see, causal heterogeneity throws a spanner in the works, because it causes predictions to fail.

---

[19] In early accounts, novelty was understood *temporally*, so a novel prediction was one that was made *before* the evidence for it was gathered (Barnes, 2018; Brush, 1994). However, temporal novelty has been criticised as being too narrow. More recent accounts view all predictions of facts that are *unknown* as novel, irrespective of the time in which they are made.

[20] An accommodation is an empirical consequence of a theory that has been verified at the time the theory is constructed.

3.2.2 Causal Heterogeneity and Prediction

How does causal heterogeneity affect predictions? The main effect of causal heterogeneity is on their accuracy. This is because predictions are based on generalisations, which, as we saw in the previous sections, are based on patterns. Scientists make predictions of the behaviour of a system based on the past behaviour of that system or the current behaviour of a similar system. If the system behaves differently with respect to its past behaviour or in comparison to similar systems, then the pattern breaks and the prediction fails. The more causal heterogeneity a system exhibits, the less likely it is that predictions will turn out to be true. To illustrate, I will examine the use of Species Distribution Models (SDMs) for predicting biological invasions.

- *Example 2. Distribution of the Emerald Ash Borer*

  The Emerald Ash Borer (EAB) is a beautiful beetle (see Figure 3D) native to East Asia, that lays its eggs under the bark of ash trees (Cuddington et al., 2018). The larvae hatch under the bark and feed on the ash tree until they become adults. Asian ash trees have a certain level of resistance to the EAB, so EAB populations in Asia are relatively small. However, EABs are highly destructive to ashes in Europe and North America. The USDA Forest Service estimates that EAB has killed hundreds of millions of ash

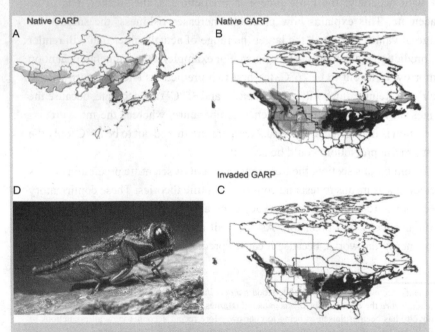

Figure 3 A, B & C GARP for Emerald Ash Borer (adapted from Sobek-Swant et al. (2012)). D Photo by U.S. Department of Agriculture, on Wikimedia Commons

trees in North America and cost hundreds of millions of dollars (USDA Forest Service, 2020).

Scientists can use SDMs, such as GARP and MAXENT, to predict the distribution of a potential invasive species in a new area (Sobek-Swant et al., 2012; Wang & Jackson, 2014). The first step is to describe the species' niche, which (in this context) refers to the environmental conditions under which the species can maintain a population. Based on the values of each parameter within the species' native distribution, the model calculates the ranges where the species can survive. The model then predicts where the species will be able to maintain populations in the new area. Figure 3 shows the native and projected distribution of the EAB using an SDM (GARP).

On the left (A) we see the model projection of the native range of the EAB. This model projection is used as a test of the accuracy of the model. If the modelled distribution corresponds to the known actual distribution of the EAB in the native range, then the scientists move on to the next step, projecting the EAB distribution in the invaded range. The top right projection (B) is the prediction of the EAB's range given the data from the native range. The bottom right projection (C) is the projection given known occurrences of EAB in North America. The overall prediction is that the EAB will likely increase its range from C to B, that is, northwards and westwards. Importantly, the scientists predicted that there would be a limit to the northern spread of EAB because of harsh winter conditions.

The underlying assumption in using SDMs for prediction is that there is a certain level of causal *homogeneity* across different systems. At first glance, this assumption makes a lot of sense as there are niche parameters that are shared between different areas. For instance, there are ash trees in East Asia and North America, while both areas have similar precipitation and temperature patterns.

Unfortunately, however, recent research has revealed that this prediction (the northern limit of EAB distribution) was inaccurate (Cuddington et al., 2018). The primary reason for this is climate change. Both the native and predicted ranges are based on actual temperature data (in this case, going back 50 years). These data sets reflected 'extreme cold events' that happened with some regularity, that is, about every six years. These events help to keep the EAB populations below certain levels. The problem is that these extreme cold events are happening less and less frequently, so the EAB populations are not being kept in check, thus allowing them to spread beyond the originally predicted range. This is an example of causal heterogeneity, as some of the causal factors that are relevant to the behaviour of the system (the frequency of extremely low temperatures) have changed.

The lesson to learn from this example is that causal heterogeneity is not just rampant, but it can be unexpected. Despite their best intentions, scientists are not always able to predict all the factors that might change so as to accommodate this change in their predictions. Surprises lead to more surprises. This points to an interesting implication of causal heterogeneity, namely that scientists are often unable to make accurate predictions, even though they are able to provide explanations of the same phenomena.

3.2.3 Explanation without Prediction?

Explanation and prediction are both important facets of scientific practice, but this has not always been so. The status of explanation and prediction and the relationship between them has changed significantly since the beginning of the twentieth century. At first, philosophers venerated prediction as the hallmark of 'true' science, because of its ability to test scientific theories. In contrast, explanation could not be characterized in precise terms, hence it was thought to be of little scientific value. At the time, there were no in-depth accounts of the nature, function and testability of scientific explanations, so it was not easily distinguishable from ordinary, every-day (and hence inherently suspect) explanations of phenomena. Many philosophers, especially those with connections to logical positivism, categorized explanation as a notion that lay beyond the realm of science, more akin to metaphysics and theology (M. H. Salmon et al., 1992). So, prediction was used to 'cleanse' explanations from mysticism, as an accurate prediction meant that the explanation of the phenomenon was scientific.

This close connection between explanation and prediction was crystalized in 1948, when Hempel and Oppenheim (1948), argued that explanation and prediction had symmetrical structure: events were logically derived from laws and various conditions, if this derivation occurred before the event took place, then it was a prediction, whereas if it took place after the event occurred, it was an explanation. In fact, according to Hempel and Oppenheim, an explanation was not fully adequate unless it could have served as the basis of a prediction of the phenomenon in question, while a failed prediction necessitated that the corresponding explanation was also a failure (Hempel & Oppenheim, 1948). In subsequent years, the 'symmetry thesis', as it became known, was slowly abandoned, as philosophers began to emphasise the value of explanation at the expense of prediction (Douglas, 2009). Nonetheless, accounts of scientific practice (espoused by philosophers and scientists) retain *vestiges* of this symmetry. As shown in Section 3.2.1, prediction is still used as a test of theory confirmation. At the same time, it is often assumed that if a theory can provide good scientific explanations, then accurate predictions ought to follow (Douglas, 2009; Strevens, 2004).

Unfortunately, this is often not the case in causally heterogeneous systems, with ephemeral patterns. Even when explanations of phenomena are perfectly adequate, predictions might still not follow. Scientists might accurately identify all the causal mechanisms that give rise to a particular phenomenon at a particular time, yet aspects of those causal mechanisms or their dynamics might change at a later date. Thus, an explanation of a particular phenomenon may be correct, in the sense that the scientists have successfully identified the causal factors that gave rise to the phenomenon. However, a prediction based on those same factors can turn out to be false, because the causal factors have changed. This is a common problem in invasion biology, where past invasions are routinely and successfully explained, yet the outcome of any particular invasion is very difficult to predict (Elliott-Graves, 2016). Consider a hypothetical but plausible example (Klironomos, 2002; Suding et al., 2013). Scientists might determine that a particular plant invasion (in area 1) occurred because of factors x, y, and z, where x is negative feedback between the native plants and the soil, which the invader is immune to, y is a particular temperature range and z is the invader's propagule pressure (the number of invasive individuals released into the invaded region (Lockwood et al., 2005)). Based on these results, scientists might predict that the same species will not succeed in invading a new area (area 2), because it is outside the optimal temperature range (in other words, y is different). However, in the next year, the temperature of the new area increases, so that it matches y, and the invasion actually succeeds. Thus, the change in the causal factors rendered the prediction inaccurate. Another way the causal factors can change is by the addition of a new causal factor that was not present in the original phenomenon and its explanation. For example, scientists might expect an invasion to succeed in another area (area 3) because all three factors (x, y and z) hold in that area. However, in addition to the negative feedback x, the invaders here are not immune to it, because of the slightly different composition of the microbiotic community of the soil. Thus, contrary to expectations, the invasion actually fails.

The upshot is that scientists investigating heterogeneous systems can make inaccurate predictions even when their explanations of the same phenomena are accurate. The next question to ask is how important is this predictive failure? In the next section, I will show that predictive failure has far-reaching implications, as it results in failed interventions. Before that, however, I will flag a small silver lining, which I will return to in Section 4. Recall that in traditional confirmatory predictions, accuracy is often too easy to achieve, hence the importance of riskiness. However, if accuracy is much less easy to achieve, then riskiness becomes less important. In other words, it is less pressing to show how predictions *could* turn out to be false when many predictions actually do turn out to be false. As we shall see, the high level of riskiness opens the door for a slightly less stringent requirement regarding precision. Reducing precision will turn out to be quite useful, as it can greatly increase the accuracy of ecological predictions and the effectiveness of interventions on ecological systems.

3.3 Effective Interventions

A substantial part of ecological research is 'applied', in the sense that it is aimed at solving real-world problems in real time, by intervening on ecological systems. Typical examples include pest control, saving a species from extinction and preventing or halting biological invasions. These interventions are based on what I call 'applied predictions': predictions of where/when an intervention will be needed and predictions of an intervention's effect on the system. Inaccurate applied predictions result in less effective or even failed interventions. A rather tragic example is the introduction of a predatory snail *Euglandina rosea* to the Pacific islands, based on the prediction that it would prey on and thus control the population of the invasive *Achatina fulica*. Unfortunately, this prediction turned out to be false, as the introduced predatory snail (*E. rosea*) preferred the native snail species to the one it was brought in to control (*A. fulica*). In fact, *E. rosea* subsequently became an invader in its own right, probably causing much more devastation on the native ecosystem than *A. fulica* (Thiengo et al., 2007). Overall, the intervention was a resounding and dramatic failure.

Ecological complexity also affects the effectiveness of interventions indirectly. The main indirect effect is *paucity of data*. Gathering data in ecology is often quite difficult. Even state-of-the-art methods are prone to patchiness and bias. Consider the task of estimating the size of a population. This is an essential step in a lot of ecological research, including most conservation efforts, yet it is exceedingly difficult and costly (Akamatsu et al., 2001; Kaschner et al., 2012; Tyne et al., 2016). The most widely used sampling method is the capture-release-recapture method, which does not work equally well for all species/populations and suffers from a number of biases (Boakes et al., 2016). For example, it is not very useful in cases with sparse data, because each capture is difficult. For species with large distribution ranges, such as marine mammals or migratory birds, the available data is often patchy, so there are gaps in the information scientists have about the population's seasonal distribution, behaviour and population dynamics (Tang et al., 2019). It can also be biased in terms of the internal structure of the population. For example, in some fish populations, the method is biased towards larger individuals, which can skew the demographic information on age-structured populations, essential in most models of population growth (Pine et al., 2003).

Sometimes ecologists rely on or supplement their own data with data collected by others. This can include historical data sets (whose quality cannot be checked) and data collected by non-experts. In some cases, such as various citizen science initiatives, these data sets prove to be extremely valuable, and of very high quality. However, in other cases, the people providing the data have incentives to distort it. This problem is especially pertinent for species that are part of commercial fisheries (e.g. cod) or species whose conservation is at odds with the interests of the fishery (e.g. dolphins), as ecologists rely on data provided by those very people with the opposing interests.

The difficulties created by ecological complexity for interventions are very important, because they have real-world consequences. In other words, they usually have *high stakes*. For example, failing to halt an invasion can incur huge economic costs in addition to the negative effects on the native ecosystem. A failure of conservation targets can lead to the extinction of a species, which is mostly irreversible. In addition, interventions are usually *time-sensitive*. Finding a solution to a problem is often not sufficient; the solution must be discovered and implemented within a particular timeframe. The scientists tasked with solving the problem must determine the extent of the danger, identify the causes of the danger, find ways to mitigate the threat and determine the best ways to implement the mitigation. All this must be achieved *before* the population drops below a certain level.

Why are these characteristics worse for applied rather than confirmatory predictions? After all, scientists making predictions for confirmation purposes don't always have optimal data sets, unlimited time or low stakes. Still, these effects tend to be much larger in the case of applied predictions. There is quite a big difference between failing to confirm a theory and failing to save a species from extinction. Similarly, time-constraints may exist in confirmatory contexts, but they are usually not as small. Poor data sets are problematic because they limit the accuracy of our best available models. For example, it is very difficult to predict the future population size after an intervention, when you cannot even determine the *current* population size with the accuracy required by the best available models. Moreover, in applied contexts, these three characteristics tend to exist simultaneously and even magnify each other. For instance, when paucity of data causes an inaccurate prediction, there is usually insufficient time to adopt an alternative method, even if such a method exists.

In order to fully appreciate how ecological complexity directly and indirectly affects predictions aimed at interventions, let us examine the case of the Yangtze River porpoise.

• *Example 3. The Yangtze River finless porpoise*

The finless porpoise (*Neophocaena asiaeorientalis asiaeorientalis*), endemic to the Yangtze river, is the world's only surviving freshwater porpoise (S.-L. Huang et al., 2017) (Figure 4). Its population has been steadily declining, and despite the existing conservation measures, in 2016 it became critically endangered. In light of this, additional, more extensive conservation measures were adopted to protect it. An essential piece of information for determining a conservation intervention is the 'time to extinction' (TE), because it affects which species are prioritised and which policies are adopted. An important paper was published in 2016, which predicted that the TE was significantly lower than previously thought (37–43 years, instead of 63 years from 2012) (Huang et al., 2017).[21]

This case bears all the hallmarks of the difficulties associated with applied predictions. The prediction here is the TE, which is used to determine the type and extent of the intervention. Typically, this prediction

(a)   (b)

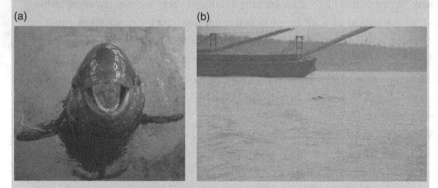

Figure 4 Yangtze finless porpoise. (A) Yangtze finless porpoise in the Institute of Hydrobiology, Chinese Academy of Sciences on Wikimedia Commons; (B) Yangtze finless porpoise in Poyang Lake, Jiangxi, China, on Wikimedia Commons

---

[21] The paper was available online from 2016.

involved high stakes. An inaccurate prediction would incur huge costs, as the porpoise population was already critically endangered, and there was little room for error. Local extinction, in this case, would mean global extinction, as the porpoise is endemic, and individuals tend not to survive in captivity. Moreover, the whole situation was extremely time-sensitive; the scientists were literally racing against the clock to determine what policies needed to be implemented or adapted before the porpoises dropped below a certain threshold.

Paucity of data was also an issue. The porpoises are extremely difficult to locate and the scientists had important gaps in their information concerning the porpoises' behavioural patterns, population dynamics and seasonal distribution (Tang et al., 2019). The scientists have to rely on sightings of dead porpoises by fishermen, though as the porpoises are most likely to die from illegal fishing methods, this data is usually biased. This patchiness or bias in the data introduces uncertainty into the models used to estimate population size, which reduces the likely accuracy of their predictions. After all, only four years before the study, the same group of scientists had predicted a significantly higher TE.[22]

Luckily, the scientists' warnings seem to have been heeded and additional conservation measures were put in place (including increased protection level of the species, increase in the size and number of protected areas, fishing bans and closer monitoring of fisheries) (J. Huang et al., 2020). These measures are considered successful, as even though the population is still decreasing, it is doing so at a much slower rate. There is now hope that the new decreased rate of decline gives us enough time to save the species from extinction, provided that the existing conservation measures are extended to increase migration between the three sub-populations (J. Huang et al., 2020; Tang et al., 2019).

This has been an admittedly pessimistic discussion of the difficulties caused by causal heterogeneity. Luckily, I believe that there are a number of options open to ecologists for conducting research that is undeniably useful and of high quality. In the next section I will examine some of these research methods and argue that they have more value for ecology than they are currently given credit for.

---

[22] There is significant overlap between the authors of the two papers. In addition, most of the authors from both papers come from the same two institutions.

4 Dealing with Ecological Complexity

It would be understandable if the discussion so far filled the reader with a degree of pessimism. I have catalogued models and theories that have failed to yield sustained generalisations and accurate predictions of ecological phenomena. I have also highlighted the implications of these failures in terms of the consequences of failed interventions. Still, I believe that this pessimism is not warranted. There are options open to ecologists, which provide a solid toolkit for dealing with ecological complexity. These are modest bottom-up generalisations, that include system-specific models and statistical tools for synthesis (such as systematic reviews and meta-analysis) and the use of models that yield imprecise predictions. Indeed, these tools are already in use by ecologists; I have not invented them. They also correspond to two of the strategies outlined in Levins's (1966) framework (type II and type III models). Yet I believe that the value of these approaches is not fully recognised as they are often deemed inadequate for the task at hand or insufficient for true progress. In this section, I will review the traditional approach (4.1) and outline two alternative approaches (4.2, 4.3) for studying complex systems, along with the main criticisms that can be levelled against them and an evaluation of these criticisms.

4.1 Reducing Complexity: The Traditional Approach

The traditional approach to dealing with complexity is to reduce it, by simplifying aspects of the phenomenon or system (Levins, 1966; Mitchell, 2003; Weisberg, 2007). In experiments, this can be achieved by omitting or controlling causal factors, whereas in models it can be achieved through abstraction (the omission of factors) and idealisation (the distortion of factors) (Elliott-Graves & Weisberg, 2014). This traditional approach works well in causally homogeneous systems, because it allows scientists to distinguish between real causal factors and noise – the real causal factors are present in many systems, whereas the noise is idiosyncratic. Thus, reducing complexity provides the additional benefit of identifying generalisations across systems. As we saw earlier, this approach is much less likely to be useful in causally heterogeneous systems, as the idiosyncratic factors of each system are often real (not mere noise) and relevant, so omitting or distorting them leads to inaccurate explanations and predictions.

Nonetheless, there are some contexts where reducing complexity can be beneficial even in causally heterogeneous systems. The first is when scientists want to identify all the instances where a particular factor is present, even when this factor is not relevant to the functioning of a particular phenomenon. This can enhance the scientists' understanding of the natural world, as it shows them which factors are common across different phenomena. This type of generalisation is often useful for pedagogical purposes as it shows how disparate phenomena might have some factors in common. For example, most populations seem to have the potential to grow exponentially, if there are no other factors curbing their growth (limited resources, competitors, predators, etc.). Of course, almost no populations grow exponentially in real-life because there are almost always multiple other factors curbing their growth. Thus, it would not be particularly useful to use an exponential growth model to predict the growth of any particular real-world population. The existence of these other factors that affect real-world populations will render the predictions of an exponential growth model inaccurate. Still, it is useful to know that most populations would grow exponentially if they could, that is, in the absence of other factors. In other words, it can be useful to know that a particular causal factor is general, even though other factors overpower it, and it does not actually manifest in real-world phenomena.

The second type of situation, where reducing complexity and applying large-scale generalisations is useful, occurs when the common causal factors are relevant for the phenomenon and its idiosyncratic factors are (i) either not relevant, (ii) are overshadowed by other factors or (iii) give rise to factors that are shared across other systems. It just so happens that in some cases scientists are lucky, so that simple, general (type I) models *do* capture the relevant causal factors of the phenomenon and actually yield accurate explanations, predictions and interventions. An example of an exceedingly simple model being used successfully can be found in the case of the Vancouver Island Marmots.

- *Example 4. Allee effect in the Vancouver Island Marmots*

  The marmots of Vancouver Island (*Marmota vancouverensis*) (Figure 5) are classified as critically endangered. It was estimated that their population had dropped 80 per cent –90 per cent since the 1980s and by the mid-2000s consisted of roughly 200 individuals (Brashares et al., 2010). The cause of this rapid decline was a mystery. The marmots were not hunted, their sources of food were unaltered, there were no new predators or competitors and the small disturbances to their habitat (small-scale logging) seemed to have a positive effect on the population, as the absence of thick tree roots in clearings made the building of burrows much easier. Nonetheless, the marmot population began to decline in the 1980s and kept on declining despite some early conservation efforts (such as the

Figure 5 Vancouver Island Marmot. Photo by Alina Fisher on Wikimedia Commons

expansion of Strathcona Provincial Park in 1995). Something had to be done quickly to preserve the species from extinction.

Brashares et al. (2010), who took on the marmot case, decided to use a simple, general (type I) model to try and determine the cause of the decline. They reasoned that, as there was no recent change in predators, competitors, disturbance to their habitat, food sources or hunting, a simple demographic model of population growth could shed light on the situation. Thus, they applied the logistic model of population growth to the Vancouver Island (V.I.) marmots (see Box 1A and Figure 6A solid line). This model describes how a population grows, given (i) its intrinsic growth rate *r* (which is the maximum possible growth rate of the population and approximately corresponds to the number of births minus the number of deaths), (ii) the density of the population and (iii) *K* the maximum size of the population that the environment can support. At first glance, the choice of model seems wrong as the growth of the V.I. marmots diverges from the model predictions (Figure 6A red line vs. Figure 6B). That is, the V.I. marmot population growth rate seems to be dropping even at low densities, in contrast to standard logistic growth.

Nonetheless, the V.I. growth rate conforms to a *recognized variation* of logistic growth, termed the 'Allee Effect' (Figure 6A dotted line), which describes populations that behave normally (i.e. according to logistic growth) at high densities, but have a positive correlation between

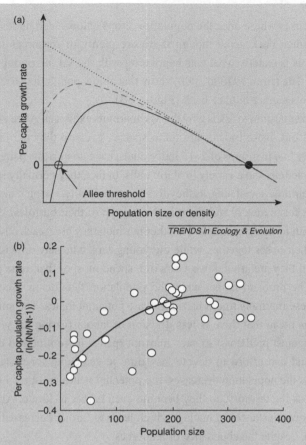

Figure 6 Logistic growth and Allee effect. (A) Per capita logistic growth (red line) and per capita logistic growth with Allee effect (dotted line). *K* is the carrying capacity (the maximum population that the environment can support). In standard logistic growth (red line) the growth rate drops as the population size increases. With an Allee effect, the population falls at low densities, despite an abundance in resources; (B) Allee effect in the V.I. marmots. The graph shows annual counts of free-living V. I. marmots 1970 to 2007. There is a clear Allee effect, a strong inverse density dependence in per capita growth rate (y axis) with respect to population size (x axis). Trend line represents least-squares quadratic fit ($R^2 = 0.55$). Data exclude animals introduced from captivity (*Figure 6B and its explanation is adapted from Brashares et al., 2010*).

the density of the population and its growth rate at low population densities (Courchamp et al., 1999). In other words, the Allee effect describes cases where once the population drops below a certain level, it cannot bounce back, even though there are plenty of resources to go around. This is exactly what was happening with the V.I. marmot population: the data (from 1970 to 2000) show that at low population sizes the V.I. marmots cannot bounce back (Figure 6B).

The next question to ask is why the V.I. marmots exhibit an Allee effect. As it turns out, these rodents are quite special, because they are highly social. They live in groups of 5–15 individuals, have an intricate pattern of social interactions and a variety of alarm calls. In fact, they normally spend most of their time socialising, as they forage communally, taking it in turns to look out for predators and share in the upkeep of their burrows. When a marmot out foraging encounters another marmot, they greet each other by touching their noses together, while the youngsters, who stay on with the family until they are around two years old, spend most of their time play-fighting. The problem is that when the population falls below a certain density, these marmots find it hard to locate potential mates. The smaller populations mean that there is less division of labour (caring for young, guarding against predators) so each marmot spends more time and effort foraging and can afford to devote less time to searching for mates. In addition, as the population decreases, the potential mates are spread more thinly across the territory, so they become even harder to locate. All this results in lower instances of mating, which leads to further decrease in the population, despite the abundance of resources.

The case of the V.I. marmots is an example of an extremely simple and general model being used to successfully explain a particular phenomenon, predict the future of behaviour of the system and form the basis of a successful intervention.[23] It is a case where the complexity of the system was significantly reduced, as the study focused only on a few demographic properties of the V.I. marmots, excluding many other factors (competitors, predators, pathogens, habitat loss, disturbance, etc.). Moreover, this model worked despite the fact that the V.I. marmots had a particular idiosyncratic characteristic that is *not shared* by other species of marmot, that is, sociality. Therefore, one may ask why I have included this example which seemingly undermines my claims that causal heterogeneity decreases generalizability.

---

[23] The sociality-induced Allee effect from this study has since been incorporated into the highly successful V.I. marmot recovery programme, which focuses predominantly on increasing the V.I. marmot population through captive breeding (as opposed to the most common alternative strategy: culling of predators) (Vancouver Island Marmot Recovery Team, 2008).

The answer is that this is a very special case – and one of the few cases where type I models are genuinely useful. It is special for two reasons. First, the complexity *could* be reduced: the factors normally operating on a population, that is, competitors, predators, pathogens, habitat loss and so on, were *not* relevant causal factors in this case. In fact, the scientists already knew that these factors were not relevant because they had remained constant before, during and after the drop in the V.I. marmot population. Second, the apparent heterogeneity of the system, that is, the uncommon sociality of the V.I. marmots, actually made this system *homogeneous* to other systems, such as plankton, plants and sessile invertebrates, because it led to *mate limitation*, which is a common cause of Allee effects in these other systems (Berec et al., 2007). In other words, this is a rare example of apparent heterogeneity resulting in causal homogeneity across different systems. As stated at the end of Section 3.1.3, the absence of causal homogeneity explains when and why some generalisations are actually possible.

Nonetheless, it is important to reiterate that this confluence of reducible complexity and effective homogeneity is far from common in ecological systems. Thus, while we should be happy when such situations occur, we should not expect them to occur frequently. More importantly, we should not criticise scientific investigations of systems that do not allow this type of generalisation. What should ecologists do then? I turn to this issue next.

4.2 Retaining Complexity: Modest Generalisations

4.2.1 Type II Models

A different approach to complexity and generalisation is employed by scientists who use type II models. Recall that type II models are those that sacrifice generality for the sake of realism and precision. Each model is built for a particular system and yields explanations and predictions for *that* system. The approach is different in the sense that the scientists employing these models and experiments value explanatory and predictive accuracy for a particular system over generalisability. Thus, the most important aspect of a type II model is to capture and represent the complexity of ecological systems. If this means that the model only applies to one system, then so be it.

Despite its potential, this approach is sometimes maligned in the literature. While authors praise the strength of the causal connections uncovered by this type of research, they lament the fact that the models do not generalise or transfer to other systems (Houlahan et al., 2017; see also discussion in

Wenger & Olden, 2012). These approaches are portrayed as *too* complex; the models they generate are considered too unwieldy and unable to promote understanding of the systems they investigate (Houlahan et al., 2017; Marquet et al., 2014). The Nobel Laureate Sydney Brenner went as far as claiming that biological research was in a state of crisis because we are 'drowning in a sea of data', and '[A]lthough many believe that more is better, history tells us that least is best' (Brenner 2012, in Marquet et al. 2014). Even though Brenner was complaining about biology in general, Marquet et al. (2014) believe that this sentiment is particularly pertinent to ecology because of the prevalence of small-scale (i.e. localised) research in the field.

I disagree. It is simply not true that scientists employing this approach are unable or unwilling to look for generalisations. For example, type II modellers might modify an existing model and investigate whether it can be transferred to a new species or area. Still, the *expectations* regarding generality are different. These scientists look for patterns, but do not always expect to find them. Moreover, even when they do find patterns, they expect them to break. In other words, the level of generality expected from this approach is usually quite constrained, as its scope is limited to variation *within* particular types of phenomena, such as disturbance (Peters et al., 2006), plant-soil feedback (Casper and Castelli 2007) or migration (Kelly and Horton, 2016).

The value of these modest generalisations can be seen in the following example of how plants cope with drought.

• *Example 5. Drought Sensitivity*

Drought is an increasingly important factor that affects all plant communities (D'Orangeville et al., 2018). As droughts are increasing in duration and frequency, it is imperative to understand and predict the strategies that plants employ to deal with drought and the likelihood of success of these strategies (Phillips et al., 2016). Strategies differ across geographical areas and taxa and depend on a number of factors such as climatic conditions (repeated short-term vs. long term water stress), plant demographic traits (age, height) and below-ground traits and interactions (root shape and size, PSF interactions) (Phillips et al., 2016). These are especially heterogeneous, so much so that most models either exclude them completely or represent them in a highly abstract way, omitting the idiosyncrasies of each case and generalises over many different cases (Phillips et al., 2016). According to Phillips et al., this reduction in complexity has resulted in models with a 'surface-bias', and a corresponding reduction in predictive accuracy.

In their paper, Phillips et al. argued for including more below-ground factors in drought sensitivity models. They pointed out that this can be achieved quite easily, as some of these models already have the capacity to include many different variables and can distinguish between fifteen different plant functional types. Moreover, the data necessary for these parameters is already available or relatively easy to collect. They argued that by adding these below-ground parameters, scientists can identify *combinations of plant traits and soil types* that confer greater sensitivity to drought. Specifically, they advocated for adding parameters for soil type and texture together with hydrology sub-models, so as to be able to predict the availability of water for a plant given a particular soil texture. In other words, with these additions to the model, scientists can determine the amount and frequency of water each type of plant needs, given the type of soil surrounding it.

Adding below-ground factors and hydrology sub-models constitutes an increase in the realism of the drought sensitivity models, which, as Phillips et al. showed, increases the models' predictive accuracy. This increase in realism comes at the expense of generality: there is no single combination of plant and soil types that confers drought tolerance to *all* plants. Each combination only works in some contexts (of soil type and availability of water). In other words, these models are flexible rather than extensively general; each time a model is applied to a particular system it includes only certain below-ground factors or dynamics –those relevant for each case.

Even though these models are not extensively general, they can identify some modest generalisations. For a start, they are only meant to apply to a subset of climatic conditions (drought) and to forests (rather than other ecosystems) –a far cry from searching for a single theory to encompass all ecological interactions. In addition, the generalisations that they generate are even more modest: they apply to particular combinations of plant and soil traits in particular contexts. Still, these extremely modest generalisations can be sufficient for scientific progress. For example, if a particular combination is useful for plants in systems with highly variable temperatures, this may provide information for how plants in a different system (with the same plant-soil trait combination) could deal with increasing temperature variability due to climate change.

The upshot is that as long as the limited scope of these generalisations is recognized, that is, their application is highly selective and carefully considered, they can be a fruitful tool for ecological research.

4.2.2 Meta-analysis

Meta-analysis is a statistical tool for analysing and synthesising the results of large numbers of individual studies (Gurevitch et al., 2018). Its primary aim is to identify causal relationships from different types of evidence (Stegenga, 2011). In some disciplines (such as ecology and evolutionary biology), meta-analyses are also used as a means for generating *generalisations* (Gurevitch et al., 2018). The aim of the meta-analysis, in this context, is to determine whether the same effect holds across different geographical locations or taxa. If causal factors are common across various different systems, then generalisation is possible. Furthermore, a meta-analysis can also reveal the limits to the scope of generalisations, by identifying the factors that cause the generalisation *not to hold* in particular cases.

The process of meta-analysis, in a nutshell, is the following: as with most scientific endeavours, a meta-analysis starts with the formulation of a *question*, for example, do alien plant species succeed in invading new areas because they lack coevolved enemies in their new ranges? (J. Parker et al., 2006). Then, the meta-analyst conducts a search of the literature to find all the papers that could be *relevant*. Next, she determines what actually is relevant for the discussion. Discarded papers usually include those that are not primary research (e.g. other meta-analyses), those that answer different questions or are beyond the scope of the meta-analysis (e.g. insect rather than plant invaders). The papers that remain are then assigned a certain *weight*, given their quality (such as sample size, transparency and depth of reported data). This data is used to calculate the *effect size*, 'a statistical parameter that can be used to compare, on the same scale, the results of different studies in which a common effect of interest has been measured' (Koricheva et al. 2013, p. 61). For example, the *effect* of herbivores on plant invasions can be measured in terms of the difference in total biomass of plants with and without herbivores. The larger the difference, the larger the effect size. Finally, the meta-analyst conducts a series of tests for biases (and some other statistical errors) and qualifies the effect size by an index of precision, such as variance, standard error or confidence interval.

How do meta-analyses contribute to generalisation in ecology? They can compare and contrast primary studies and identify where causal relationships hold and where they do not, thus revealing the limits of a generalisation's scope. Sometimes, this process can even yield new insights about ecological phenomena. I will illustrate with the case of the Enemy Release Hypothesis in invasion biology.

• *Example 6. The Enemy Release Hypothesis:*

Invasive species around the globe pose an important threat to native natural and agricultural communities and incur huge costs for their prevention and management. Despite their importance, scientists have yet to uncover a general theory of invasion (Elliott-Graves, 2016). Part of the problem seems to be that there are numerous theories for why certain invasions succeed and others fail, and each theory has both empirical data that support and also that contradict it. (Jeschke et al., 2012). One of these hypotheses is the 'enemy release hypothesis' (ERH) (Heger & Jeschke, 2014). The basic idea is quite simple: alien species can thrive in new territories because they do not encounter their traditional enemies. In the case of plant invasions (where the hypothesis has gained the most traction), the ERH predicts that native herbivores are less likely or able to consume alien plants, thus giving alien plants a competitive advantage over native plants.

There are observational and experimental results that confirm the ERH. For example, there are documented cases where alien plants or their seeds are not consumed by native predators (J. Parker et al., 2006). However, there are also experimental and observational results that contradict the ERH. For example, there are a number of cases where alien plants actually attract native herbivores and this effect is significant enough to reduce the alien plants' seed production and survival (J. Parker et al., 2006). Studies conducted at different scales also seem to produce different results: large-scale biogeographical analyses tend to show a reduction in the diversity of enemies in the introduced range compared with the native range, whereas medium to small scale community studies tend to find that alien species are no less affected by enemies than native species in the invaded community (Colautti & MacIsaac, 2004).

Parker et al. (2006) decided to review the existing literature on the ERH and conducted a meta-analysis to see if they could settle the matter. What could account for the contradictory findings? They found two interesting results. First, there were some cases where native herbivores decreased the abundance of alien plants. Second, there were cases where alien herbivores increased the abundance of alien plants. Both of these results contradict the ERH, as they show that alien plants *are* subject to predation in new areas (i.e. they acquire new enemies) and their *old enemies* actually help them increase in abundance, rather than hindering them.

Parker et al. continued analysing the data to see if there were any other similarities and differences between the primary studies. They noticed three things: first, the negative effect of *native* herbivores on the alien plants was weaker than the positive effect of *alien* herbivores on them (28 per cent reduction in the former vs 65 per cent increase in the latter). Second, some studies focused on *invertebrate* herbivores while others focused on *vertebrates*. These two categories also had consistent differences in effects: native *vertebrate* herbivores had a three to five-fold larger negative impact on alien plant survival than native *invertebrate* herbivores. Third, all the alien herbivores in these studies were *generalists* (they prey indiscriminately on many species of plant). This is important because we know that vertebrate herbivores tend to be generalists, while invertebrate herbivores tend to be *specialists* (i.e. they prey on specific plant species).

If we put all these points together, a clearer picture of the scope of ERH emerges. When the native herbivores are specialists and there are no alien herbivores, the ERH works as expected: the alien plants are released from their old enemies and the native specialists do not bother them, as they continue to focus on their preferred native plants. But the ERH does not work as expected in the following cases: (i) when the native herbivores are generalists, they prey on native and alien plants, so there is no significant enemy release effect. (ii) when alien generalist herbivores are also present, things become more complicated. The alien plants might still have an ERH effect with respect to their native specialist herbivores. However, this effect is overshadowed by the effect of alien generalist herbivores (their old enemies). These generalists might actually prefer the native plants, so that they consume more of the natives than the alien plants, giving the alien plants a competitive advantage over the native plants.

Interestingly, enemy release is also part of the explanation of points (i) and (ii), though from an entirely different perspective. So far, we have been thinking of predators as the enemies of plants, but plants can also have negative effects on predators, through co-evolution. Plants from the same regions as herbivores often evolve mechanisms to deter these herbivores. However, these mechanisms are not very effective; the herbivores will still consume them, and deal with the consequences (e.g. by spending significant amounts of time and energy digesting their food and neutralising toxins). Still, the same generalists will preferentially consume plants that have not evolved these defence mechanisms if they are given the chance. This is exactly the chance given to them in the new regions, where they preferentially consume the native plants. Indeed, the meta-analysis found strong support for this hypothesis, as 88 per cent of alien plant species in the studies shared the same native ranges as the alien generalist herbivores.

In fact, these insights were also able to explain another perplexing phenomenon, namely why it is much more common for European plants to invade areas outside Europe, rather than vice versa. The answer is that generalist herbivores from Europe, such as pigs, horses and cattle, are more widespread than generalist herbivores from other continents and contribute more often to the success of exotic plants with which they have co-evolved.

To sum up, the meta-analysis teaches us two different things about the scope of the ERH. First, the notion of ERH, as it stands, is too broad. We need to specify different contexts for who is the enemy in each case, and what type of enemy it is. An original suggestion for how to achieve this further specification with respect to the ERH and to other theories can be found in a series of papers by Tina Heger and colleagues (Bartram & Jeschke, 2019; Heger et al., 2021; Heger & Jeschke, 2014). Second, having to constrain the scope of the ERH did not incur huge costs, nor was it detrimental to our understanding of invasion biology. Examining when, where and why the ERH does not hold provided information about the hypothesis and about invasions more generally. In fact, this whole process of constraining the scope of the ERH was instrumental in providing an answer to a question that had long perplexed ecologists – the differential frequency of plant invasions out of versus into Europe. In short, if the ecological complexity of a phenomenon 'forces' scientists to restrict the scope of a generalisation, this should not be seen as a failure in ecological research, but as an opportunity to acquire knowledge about the systems and phenomena under investigation.

Before I turn to the next tool at ecologists' disposal, there is a general lesson to be learned from this approach to generalising in ecology. The types of generalisations I have discussed have an important point in common: they probe, explore and test *whether* a generalisation is possible, rather than assuming it is possible. Observing a pattern, for these scientists, means that it could result in a useful generalisation, not that it automatically entails a generalisation. This may seem like a small point, but it is actually very important. Changing our attitude and expectations towards generalisations, is a key step in recognising the potential of generalisations in ecology. We should not be dismissive or embarrassed when generalisations turn out to be limited in scope or break down completely. We should not even expect many generalisations to actually exist in complex, heterogeneous systems. This does not mean that we should stop trying to identify them. It means that we should value them for what they are when we do find them, and that we should also value the research that does not aim for generality or achieves generality at the expense of precision – this is the research I turn to next.

4.3 Incorporating Uncertainty: Imprecise Predictions

The most controversial strategy for ecological research is to incorporate uncertainty in models by sacrificing the *precision* of their predictions. In general terms, imprecise predictions are those that refer to the existence of a phenomenon or effect, without specifying its extent or magnitude. For example, a model could predict a trend, such as 'population x will increase', without specifying by how much it will increase. Some predictions are imprecise because they predict a range of values for a particular variable. For example, the temperature in 2030 will increase by 1.5–3.5 degrees. Other predictions are imprecise because they include imprecise probabilities; for example, there is 0.6–0.8 probability that organism y will be extinct in twenty years.

It should be clear from these examples and from the discussion in Section 3.2, that precision is not a binary characteristic, but comes in degrees. In fact, imprecision is a way of representing *uncertainty* (Elliott-Graves, 2020a; W. Parker, 2010; W. Parker & Risbey, 2015; Smith & Stern, 2011). That is, scientists can use imprecision to convey that they are not certain about a model's predictions. The more uncertain the result, the less precise the prediction. Table 2 shows predictions for variable x in decreasing levels of precision (adapted from Parker & Risbey 2015).

In ecology, imprecision usually appears as the output (predictions) of models. The imprecision in these models usually corresponds to levels (c)–(e) in Figure 6. The models themselves correspond to Levins's type III strategy (Section 2.2.1). The main attribute of these models is that they can achieve generality without sacrificing realism. By representing each factor imprecisely, they can include many more factors than a type I model. They are thus said to contain *idealisations of specificity* rather than *idealisations of veracity* (Justus 2006, p. 659). That is, imprecise models represent systems *veridically* (without omitting or distorting most of their variables). In contrast, maximally precise models employ different methods of simplification, that is, making unrealistic assumptions and/or omitting causal factors altogether, which decreases the 'veracity' of the model. This type of idealisation often mischaracterizes *salient features* of systems, resulting in inaccurate explanations and/or predictions (Justus, 2005, 2006).

Table 2 Prediction and uncertainty

	Prediction of x	Interpretation of the associated uncertainty
(a)	x will increase by 3 units (0.95 probability)	This prediction is virtually certain. The scientist has provided a precise point probability for the value of x.
(b)	x will increase by 3 units (0.6–0.8 probability)	This prediction is imprecise because it involves imprecise probabilities, i.e. a range. It is also possible to express this level of uncertainty with a qualitative confidence level, such as 'medium'
(c)	x will increase by 2–5 units (0.95 probability)	This prediction is even more imprecise. Even though the scientist has provided a precise probability estimate, the value for the increase of x is a range.
(d)	x will increase by 2–5 units (0.6 –0.8 probability)	Here precision decreases even more, as both the value of x and the probability of the prediction are expressed as ranges.
(e)	x will increase by more than 2 units	Here the precision has decreased further, as the value of x is expressed as an order of magnitude estimate.
(f)	x will increase	The levels of uncertainty are so high, that the value of x can only be expressed as a trend.
(g)	the future value of x is entirely unknown	There is so much uncertainty with respect to x than no prediction is possible.

Despite these advantages, imprecise predictions have traditionally been viewed with suspicion in philosophy of science, for two main reasons.[24] First, they are thought to be insufficiently *risky* to provide strong enough tests for theory confirmation (Orzack & Sober, 1993; Rosenberg, 1989, see also Section 3.2). If predictions are the primary method by which we test and choose between theories, then they must not be too easy to obtain. The worry is that as precision trades off with accuracy, the less precise the prediction, the more likely it is to be accurate. Thus, if we sacrifice too much precision, the predictions will not be able to discriminate between higher and lower quality theories. The second criticism of imprecise predictions is that they could lead to ineffective, insufficient or inappropriate interventions (Houlahan et al., 2017; Orzack & Sober, 1993; Rosenberg, 1989). For example, if scientists are trying to save a particular population x from extinction, they might use a model that predicts that population's qualitative trend after an intervention: a drop in the predator population y will cause the population of the prey x to rise. The scientists will then intervene to reduce population y. The worry is that if the scientists do not know by *how much* the predator population is expected to fall with the intervention, their intervention could be ineffective. They could reduce the predator population y, but not reduce it enough, which would not save species x from extinction. Alternatively, they might reduce the predator population too much, spending scarce resources that could be put to better use. Thus, the story goes, it is better to have specific information about how much the population of x will rise given a specific intervention on population y.

These criticisms might seem plausible, but they are not warranted. First, it should be noted that imprecise model predictions are not exceedingly imprecise; that is, they do not correspond to levels (f) and (g) in Table 2. Rather, they correspond to levels (c) and (d), and occasionally level (e). That is, there are many values that they could take which would render them false, so they are not altogether without risk. Second, in causally heterogeneous systems, predictive accuracy is much more difficult to achieve than the critics presuppose. It is not the case that most predictions turn out to be true, so that we need stronger tests for our theories. Rather, as explained above and in the previous section, predictions in ecology tend to come out false quite often. In these cases, even an imprecise but accurate prediction is better than a number of precise but wildly inaccurate predictions.

Third, imprecise (type III) models are often better than both type I and type II models at dealing with biased, patchy or low-quality data (see Section 3.3). This

---

[24] Some critiques of imprecise predictions claim that they are opaque, mathematically unsound or untrustworthy (Gonzalez, 2015). However, this criticism does not apply to imprecise predictions in ecology, because they are the outputs of mathematical models.

is because poor data sets lead to errors in the estimation of parameters, which result in errors in the model's output. Often (especially when the models have large numbers of parameters), these errors can persist despite statistical methods used to reduce or correct them (Dambacher et al., 2003; Novak et al., 2011). In contrast, type III models are built for imprecise parameter estimates, so even though their predictions are imprecise, the entire range that the prediction spans, tends to be more accurate. In other words, the predictions of precise models are often further away from the actual value than the entire range of the imprecise model prediction. For example, Novak et al. (2011) found that even if there is observational and experimental data on each particular species within an eco-system but insufficient data on the indirect effects of each population on other populations within the ecosystem, the predictive accuracy of highly precise models dwindles rapidly (e.g. to about the level of flipping a coin).

In order to fully appreciate the value and potential of imprecise models, one only has to look into the case of the kōkako. The scientists here used two imprecise models to bring back the kōkako from the brink of extinction. Importantly, these models yielded predictions that were sufficient in terms of both riskiness and for formulating and implementing a successful intervention.

• *Example 6 – Saving the kōkako from extinction*

The North Island kōkako is a bird endemic to New Zealand (see Figure 7). In 1999 the species was reduced to 400 pairs, because of predation from the 'unholy trinity of pestilence': possums, stoats and rats (Hansford, 2016) (Figure 7). The scientists tasked with saving the kōkako from extinction faced three difficulties (Ramsey & Veltman, 2005):

(i) Limited resources. Ideally, the scientists would have simply eradi-cated all three predator populations, but this was not feasible. Resources only allowed for a reduction of predator populations.

(ii) Limited time. As the kōkako population was already so low, the conservation strategy needed to be conceptualised and implemented quickly – before the kōkako population fell to a level from which it could not be restored.

(iii) Poor/patchy data. It was practically impossible to determine the exact size of each population and consequently the precise rates of competi-tion and predation within the community.

The conservation strategy was based on a combination of two *impre-cise models* (Ramsey & Veltman 2005). The first, *loop analysis* (Box 1 C I), was used to identify all the dynamic relationships between

Figure 7 North Island kōkako and predators. (A) Photo by Doug Mak on Wikimedia Commons; (B) Photo from Hansford (2016) Radio New Zealand

populations in the community, along with their *quality* (positive (1), neutral (0), negative (−1)). Scientists can use the conjunction of qualitative effects to determine which predator population should be the focus of an intervention so that the prey population is saved from extinction. In this case, loop analysis predicted that an increase in all predator populations would have a negative effect on the kōkako population, but that an increase in the rat population would have a stronger effect than the other two predators because it would result in a net total of two negative feedback cycles (one from predation and one from competition) (Box 1 C II).

The second model was a 'fuzzy interaction web' (FIW). FIWs take imprecise information on the size of populations within an ecological community, such as population abundance data that is incomplete, and create 'fuzzy sets'. Recall that precise and accurate data on population sizes is impossible to attain. With FIWs, however, scientists have a way to get around this problem. They can conduct the entire investigation without knowing the exact size of any population in the community. They do this by measuring 'tracking rates', that is, the percentage of traps that are full on any given night. These are used to determine what counts as low, medium or high abundance for this community. Thus, for example, high abundance could occur when 20 per cent or above of the traps each night are full and low abundance could occur when less than 10 per cent of traps are full. Values between 10 per cent and 20 per cent could mean 'quite high', 'moderately high', 'quite low' and so on (Figure 8).

The 'fuzziness' is a way of representing this uncertainty about population sizes. The concept comes from 'fuzzy logic', an approach where a variable can take multiple values. In classical logic, conclusions are either true or false, and each variable either is or is not a member of a set.

For example, a monochromatic marble is either blue or it is not blue; it cannot belong and not belong to the set of blue things. However, in fuzzy logic, the boundaries between sets are not 'crisp', so membership of a set is *partial*. This does not mean that the marble is and is not blue at the same time, but that we are not certain about whether the marble is or is not blue. In the case of the kōkako, the fuzziness represents the fact that the scientists are not 100 per cent certain about where the boundaries between the categories of low, medium and high are. Figure 8 shows the fuzzy set membership for each of the species in the community.

The scientists can use these fuzzy sets to make predictions about the effect of each population on the other populations in the community. Figure 9 summarises the predictions yielded by the FIW. It shows that controlling all three populations would have a high effect on the kōkako population, whereas controlling rats and possums would have a moderate effect on the kōkako population. As a moderate effect is sufficient for bringing back the kōkako population to acceptable levels, the most efficient intervention should focus on the rats and possums, and not worry about the stoats. More specifically, the FIW predicted that the rats need to be kept at a tracking rate of 'about 11 per cent or lower' and possums at a tracking rate 'below about 10 per cent', so that the kōkako population is maintained at 'moderate levels' (Ramsey & Veltman 2005, p. 914).

How does this example hold up to the criticisms of imprecise models? Recall that the first criticism was that imprecise models are not sufficiently *risky*. A non-risky prediction, in this context, would be that *all* predator populations should be controlled. The prediction is not risky because it would be satisfied if the desired effect was achieved by culling any one, two or all three predator populations. However, neither loop analysis nor the FIWs made that *sole* prediction. While both models yielded the prediction that culling all three predators would have a positive effect on the kōkako population, they also made the much riskier prediction that this extensive intervention was unnecessary, and that the intervention could focus on just the rat population (loop analysis) or the rat and possum population (FIW).

The second criticism was that interventions based on imprecise models are ineffective. Yet the kōkako case is a perfect example of an effective intervention. The populations have indeed increased, and additional populations have been established. Their range has expanded to twenty-two different sites. In fact, the kōkako is now the poster child of successful conservation in New Zealand. But how do we know that these models were influential in the actual intervention? As it turns out,

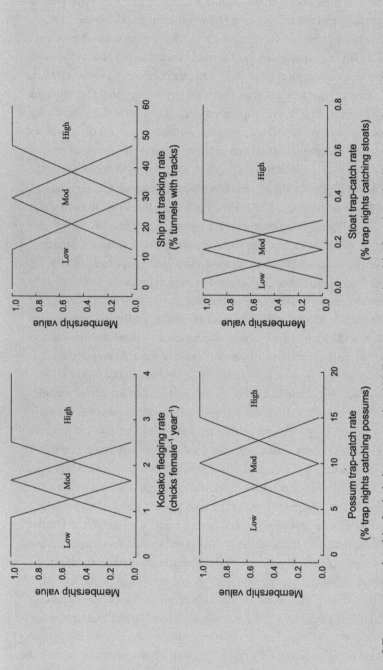

Figure 8 Fuzzy set membership for kōkako community. (A) Fuzzy set membership functions for linguistic descriptions of species abundances ('low', '(mod)erate', 'high') for each of the animal species in the kōkako Fuzzy Interaction Web (reprinted from Ramsey & Veltman 2005, supplementary materials)

Species controlled	Population abundance of rats	Population abundance of possums	Population abundance of stoats	Kokako fledging rate
Rats	Low	Mod	Low	Low
Possums	Mod	Low	Mod	Low
Stoats	Mod	Mod	Low	Low
Rats and possums	Low	Low	Mod	Mod
Rats and stoats	Low	Mod	Low	Low
Possums and stoats	Mod	Low	Low	Low
All three	Low	Low	Low	High

Figure 9 Kōkako fuzzy interaction web predictions. Imprecise predictions of the magnitude of the effect on the equilibrium kōkako fledgling rate resulting from sustained single and multispecies control of nest predators from the FIW 'trained' model (reprinted from *Ramsey & Veltman 2005*)

one of the paper's authors (Veltman), was based at the Science and Research Unit of the New Zealand Department of Conservation, which was in charge of the kōkako conservation project. In addition, other papers she (co)authored include research on direct and indirect effects of pest (rat, stoat and possum) control on ecological communities (Tompkins & Veltman, 2006). This work was part of a larger project within the Department of Conservation, contributing to a paper on imprecise models for deer control (also a pest in New Zealand) (Ramsey et al., 2012). Finally, subsequent publications of the Department of Conservation incorporate the research and recommendations from papers by these scientists (see for example, Brown et al., 2015). So, it is extremely likely that these model predictions were actually used in the intervention on the kōkako and were instrumental in saving the population from extinction.

I hope that the discussion in this section has dispelled any pessimism regarding the ability of scientists to deal with ecological complexity in the systems they investigate. I have outlined examples from three distinct types of investigation, all of which allow ecologists to conduct high quality research and can form the basis of successful interventions. These types of research are (i) type II models (that yield modest generalisations), (ii) statistical tools such as meta-analysis (that give scientists the means to test and probe the scope of existing generalisations), and (iii) imprecise (type III) models that do not suffer from the limitations of the other models in cases involving poor and patchy data. I should note that my claim is not that these are the only tools that ecologists should use. I do not think that ecologists should drop all other types of inquiry and focus exclusively on these three types of research. I am also not arguing for the complete elimination of the traditional approach (exemplified by simple, general, type I models). There will undoubtedly be cases, such as the case of the V.I. marmots, where type I models are likely to be successful and should therefore be preferred. The point is that these cases are not the norm, but quite special. Their frequency is much lower than the advocates of the traditional approach seem to expect. In contrast, the situations where the other approaches outlined in this section will be useful are going to be much more frequent in complex, heterogeneous systems. Therefore, we should not be surprised, much less embarrassed, when an ecological investigation calls for type II or type III models, or when the only generalisations that can be made are quite modest.

5 Concluding Remarks

Causal heterogeneity is an underappreciated aspect of complexity. both in terms of how interesting it is and in terms of the extent of its effect on scientific practice. Failing to include causal heterogeneity in the concept of complexity obscures what makes ecological systems interesting but also so difficult to study. The aim of this Element was to put causal heterogeneity on the map as an integral aspect of ecological complexity. I showed that complexity *sans* heterogeneity cannot account for the extent and frequency of the epistemic difficulties ecologists face. I argued that causal heterogeneity magnifies the Levinsian trade-offs between the desiderata of generality, realism and precision. I then examined the epistemic difficulties in more detail, explaining how and why causal heterogeneity causes generalisations to fail, predictions to be inaccurate and interventions to be ineffective, illustrating with examples from recent ecological research. In the final section, I outlined three alternative strategies for dealing with ecological complexity. These strategies are already in use but undervalued by ecologists and philosophers. Instead of lamenting the need to use these strategies, we should recognise their potential for providing insights into the workings of these interesting, intricate and complicated systems.

A key motivation for my engaging with the topic of ecological complexity is the periodic but persistent worry shared by numerous ecologists that their discipline is somehow not up to scratch. In the introduction, I catalogued a number of papers whose authors bemoan the lack of extensive generalisations and predictions that are simultaneously accurate and highly precise. My worry, on the other hand, is that these ecologists are relying on outdated philosophy of science.[25] Just as Sandra Mitchell argued that the complexity of biological systems should force us to update our notion of laws rather than question the maturity of the discipline of biology, so I argued that the causal heterogeneity of ecological systems should force us to update our notion of generalisability and prediction. Ecology is not an immature science; it is a difficult science. Ecological systems are surprising, irregular and mercurial. The scientists who manage to get information about them should be commended in their endeavours, even if they suffer setbacks and complications.

Of course, this does not mean that all ecological research is of equally high quality, as in every discipline there can be cases of bad science. The point is how we determine what counts as good or bad science. I have argued that when you have complex and heterogeneous systems, the traditional tools for determining what counts as successful research will not necessarily yield accurate results. Modest generalisations and imprecise predictions are likely to be more accurate than their more general and precise counterparts. Thus, an abundance of this type of research should not be seen as a limitation of ecology, but as a feature. These are legitimate ways in which ecologists can understand and deal with ecological complexity.

---

[25] If the term outdated is too strong for some readers, it can be replaced with 'inapt for the purposes of ecology' – the argument remains unchanged.

Bibliography

Akamatsu, T., Wang, D., Wang, K., & Wei, Z. (2001). Comparison between visual and passive acoustic detection of finless porpoises in the Yangtze River, China. *The Journal of the Acoustical Society of America, 109*(4), 1723–7.

Anand, M., & Orlóci, L. (1996). Complexity in plant communities: The notion and quantification. *Journal of Theoretical Biology, 179*, 179–86.

Arthur, W.B. (1999). The End of Certainty in Economics. In: Aerts, D., Broekaert, J., Mathijs, E. (eds) Einstein Meets Magritte: An Interdisciplinary Reflection. Einstein Meets Magritte: An Interdisciplinary Reflection on Science, Nature, Art, Human Action and Society, vol 1. Springer, Dordrecht. https://doi.org/10.1007/978-94-011-4704-0_14

Barkai, A., & McQuaid, C. (1988). Predator-prey role reversal in a marine benthic ecosystem. *Science, 242*(4875), 62–4.

Barnes, E. C. (2018). Prediction versus accommodation. In E. N. Zalta (Ed.), *The Stanford Encyclopedia of Philosophy.* https://plato.stanford.edu/archives/fall2018/entries/prediction-accommodation/

Barrett, J., & Stanford, P. K. (2006). Prediction. In J. Pfeifer & S. Sarkar (Eds.), *The Philosophy of Science: An Encyclopedia* (pp. 585–599). Routledge.

Bartram, I., & Jeschke, J. (2019). Do cancer stem cells exist? A pilot study combining a systematic review with the hierarchy-of-hypotheses approach. *PLoS One, 14*(12), e0225898.

Barua, M. (2011). Mobilizing metaphors: The popular use of keystone, flagship and umbrella species concepts. *Biodiversity & Conservation, 20*(7), 1427–40.

Bascompte, J., & Solé, R. (1995). Rethinking complexity: Modelling spatio-temporal dynamics in ecology. *Trends in Ecology & Evolution, 10*(9), 361–6.

Beck, M. W. (1997). Inference and generality in ecology: Current problems and an experimental solution. *Oikos, 78*(2), 265–73.

Beckage, B., Gross, L. J., & Kauffman, S. (2011). The limits to prediction in ecological systems. *Ecosphere, 2*(11), 1–12.

Benincà, E., Huisman, J., Heerkloss, R. et al. (2008). Chaos in a long-term experiment with a plankton community. *Nature, 451*(7180), 822–5.

Berec, L., Angulo, E., & Courchamp, F. (2007). Multiple Allee effects and population management. *Trends in Ecology & Evolution, 22*(4), 185–91.

Bishop, R. C. (2011). *Metaphysical and Epistemological Issues in Complex Systems* (C. Hooker, Ed.; Vol. 10, pp. 105–136). Elsevier.

Boakes, E. H., Fuller, R. A., McGowan, P. J. K., & Mace, G. M. (2016). Uncertainty in identifying local extinctions: The distribution of missing data and its effects on biodiversity measures. *Biology Letters, 12*(3), 20150824.

Bonenfant, C., Gaillard, J. M., Coulson, T. et al. (2009). Empirical evidence of density-dependence in populations of large herbivores. *Advances in Ecological Research, 41*, 313–57.

Brashares, J. S., Werner, J. R., & Sinclair, A. R. E. (2010). 'Social meltdown' in the demise of an island endemic: Allee effects and the Vancouver Island marmot. *Journal of Animal Ecology, 79*(5), 967–53.

Brown, K., Elliott, J. I., & Kemp, J. (2015). *Ship rat, stoat and possum control on mainland New Zealand* [Scientific Report]. New Zealand Department of Conservation.

Brush, S. G. (1994). Dynamics of theory change: The role of predictions. *PSA: Proceedings of the Biennial Meeting of the Philosophy of Science Association, 1994*(2), 133–45.

Cartwright, N. (1989). *Nature's Capacities and Their Measurement.* Oxford University Press.

Casper, B. B., & Castelli, J. P. (2007). Evaluating plant-soil feedback together with competition in a serpentine grassland. *Ecology Letters, 10*(5), 394–400.

Clark, A. T., Ann Turnbull, L., Tredennick, A. et al. (2019). Predicting species abundances in a grassland biodiversity experiment: Trade-offs between model complexity and generality. *Journal of Ecology, 108*(2), 774–87.

Clarkson, K., Eden, S. F., Sutherland, W. J., & Houston, A. I. (1986). Density dependence and magpie food hoarding. *Journal of Animal Ecology, 55*(1), 111–21.

Colautti, R. I., & MacIsaac, H. J. (2004). A neutral terminology to define 'invasive' species. *Diversity and Distributions, 10*(2), 135–41.

Colyvan, M., & Ginzburg, L. R. (2003). Laws of nature and laws of ecology. *Oikos, 101*(3), 649–53.

Comita, L. S., Muller-Landau, H. C., Aguilar, S., & Hubbell, S. P. (2010). Asymmetric density dependence shapes species abundances in a tropical tree community. *Science, 329*(5989), 330–2.

Cooper, G. (1998). Generalizations in ecology: A philosophical taxonomy. *Biology & Philosophy, 13*(4), 555–86.

Cottee-Jones, H. E. W., & Whittaker, R. J. (2012). Perspective: The keystone species concept: A critical appraisal. *Frontiers of Biogeography, 4*(3), 117–27.

Courchamp, F., Clutton-Brock, T., & Grenfell, B. (1999). Inverse density dependence and the Allee effect. *Trends in Ecology & Evolution, 14*(10), 405–10.

Courchamp, F., Dunne, J.A., Le Maho, Y., et al. (2015). Fundamental ecology is fundamental. *Trends in Ecology and Evolution, 30*, 9–16.

Cuddington, K., Sobek-Swant, S., Crosthwaite, J. C., Lyons, D. B., & Sinclair, B. J. (2018). Probability of emerald ash borer impact for Canadian cities and North America: A mechanistic model. *Biological Invasions, 20*(9), 2661–77.

Dambacher, J. M., Li, H. W., & Rossignol, P. A. (2003). Qualitative predictions in model ecosystems. *Ecological Modelling, 161*(1–2), 79–93.

Doak, D. F., Estes, J. A., Halpern, B. S. et al. (2008). Understanding and predicting ecological dynamics: Are major surprises inevitable? *Ecology, 89*(4), 952–61.

Donohue, I., Hillebrand, H., Montoya, J. M. et al. (2016). Navigating the complexity of ecological stability. *Ecology Letters, 19*(9), 1172–85.

D'Orangeville, L., Maxwell, J., Kneeshaw, D. et al. (2018). Drought timing and local climate determine the sensitivity of eastern temperate forests to drought. *Global Change Biology, 24*(6), 2339–51.

Douglas, H. (2009). Reintroducing prediction to explanation. *Philosophy of Science, 76*(4), 444–63.

Douglas, H., & Magnus, P. D. (2013). State of the field: Why novel prediction matters. *Studies in History and Philosophy of Science, 44*(4), 580–9.

Egler, F. E. (1986). 'Physics envy' in ecology. *Bulletin of the Ecological Society of America, 67*(3), 233–5.

Elliott-Graves, A. (2016). The problem of prediction in invasion biology. *Biology & Philosophy, 31*(3), 373–93.

Elliott-Graves, A. (2018). Generality and causal Interdependence in ecology. *Philosophy of Science, 85*(1), 1102–114.

Elliott-Graves, A. (2020a). The value of imprecise prediction. *Philosophy, Theory, and Practice in Biology, 12* (4).

Elliott-Graves, A. (2020b). What is a target system? *Biology & Philosophy, 35*(2), 1–28.

Elliott-Graves, A. (2022). What are general models about? *European Journal for Philosophy of Science, 12*(74).

Elliott-Graves, A., & Weisberg, M. (2014). Idealization. *Philosophy Compass, 9*(3), 176–85.

Fahrig, Lenore, 'Forty years of bias in habitat fragmentation research', in Peter Kareiva, Michelle Marvier, and Brian Silliman (eds), *Effective Conservation Science: Data Not Dogma* (Oxford, 2017; online edn, Oxford Academic, 21 Dec. 2017), https://doi.org/10.1093/oso/9780198808978.003.0005

Fischer, R., Rodig, E., & Huth, A. (2018). Consequences of a reduced number of plant functional types for the simulation of forest productivity. *Forests, 9*(8), 460.

Frigg, R. (2009). Models and fiction. *Synthese, 172*(2), 251–68.

Giere, R. N. (2004). How models are used to represent reality. *Philosophy of science, 71*(5), 742–52.

Godfrey-Smith, P. (2006). The strategy of model-based science. *Biology & Philosophy, 21*(5), 725–40.

Gonzalez, W. J. (2015). *Characterization of Scientific Prediction and Its Kinds in Economics* (Vol. 50, pp. 47–76). Springer International.

Grace, J. (2019). Has ecology grown up? *Plant Ecology & Diversity*, *12*(5), 387–405.

Gurevitch, J. et al., (2018). Meta-analysis and the science of research synthesis. *Nature*, *555*(7695), 175–82.

Hansford, D. (2016). *War on pests avoids targeting pets and dinner*. Radio New Zealand. www.rnz.co.nz/news/on-the-inside/309756/war-on-pests-avoids-targetting-pets-and-dinner.

Heger, T., Aguilar, C., Bartram, I. et al. (2021). The hierarchy-of-hypotheses approach: A synthesis method for enhancing theory development in ecology and evolution. *BioScience*, *71*(4), 337–49.

Heger, T., & Jeschke, J. (2014). The enemy release hypesis as a hierarchy of hypotheses. *Oikos*, *123*(6), 741–50. https://doi.org/10.1111/j.1600-0706.2013.01263.x.

Hempel, C. G., & Oppenheim, P. (1948). Studies in the logic of explanation. *Philosophy of Science*, *15*(2), 135–75.

Hitchcock, C., & Sober, E. (2004). Prediction versus accommodation and the risk of overfitting. *British Journal for the Philosophy of Science*, *55*, 1–34.

Hooker, C. (Ed.). (2011). *Philosophy of Complex Systems* (Vol. 10). Elsevier.

Hooper, D. U., Chapin III, F. S., Ewel, J. J., et al. (2005). Effects of biodiversity on ecosystem functioning: a consensus of current knowledge. *Ecological Monographs*, *75*(1), 3–35.

Houlahan, J., McKinney, S., Anderson, M., & McGill, B. (2017). The priority of prediction in ecological understanding. *Oikos*, *126*(1), 1–7.

Huang, J., Mei, Z., Chen, M. et al. (2020). Population survey showing hope for population recovery of the critically endangered Yangtze finless porpoise. *Biological Conservation*, *241*, 108315.

Huang, S.-L., Mei, Z., Hao, Y. et al. (2017). Saving the Yangtze finless porpoise: Time is rapidly running out. *Biological Conservation*, *210*, 40–6.

Jeschke, J., Gómez Aparicio, L., Haider, S. et al. (2012). Support for major hypotheses in invasion biology is uneven and declining. *NeoBiota*, *14*(0), 1–20.

Johnson, K. (2007). Natural history as stamp collecting: A Brief History. *Archives of Natural History*, *34*(2), 244–58.

Jones, M. R. (2005). *Idealization and Abstraction: A Framework*. (M. R. Jones & N. Cartwright, Eds., pp. 173–217). Rodopi.

Justus, J. (2005). Qualitative scientific modeling and loop analysis. *Philosophy of Science*, *72*(5), 1272–86.

Justus, J. (2006). Loop analysis and qualitative modeling: Limitations and merits. *Biology & Philosophy*, *21*(5), 647–66.

Justus, J. (2021). *The Philosophy of Ecology: An Introduction*. Cambridge University Press.

Kaschner, K., Quick, N. J., Jewell, R., Williams, R., & Harris, C. M. (2012). Global coverage of cetacean line-transect surveys: Status quo, data gaps and future challenges. *PLoS One*, *7*(9).

Kaunisto, S., Ferguson, L. V., & Sinclair, B. J. (2016). Can we predict the effects of multiple stressors on insects in a changing climate? *Current Opinion in Insect Science*, *17*, 55–61.

Kelly, J. F., & Horton, K. G. (2016). Toward a predictive macrosystems framework for migration ecology. *Global Ecology and Biogeography*, *25*(10), 1159–65.

Kingsland, S. (1995). *Modeling Nature*. University of Chicago Press.

Kingsland, S. (2005). *The Evolution of American Ecology, 1890–2000*. JHU Press.

Kitcher, P. (1981). Explanatory unification. *Philosophy of Science*, *48*(4), 507–31.

Klironomos, J. N. (2002). Feedback with soil biota contributes to plant rarity and invasiveness in communities. *Nature*, *417*(6884), 67–70.

Knuuttila, T., & Loettgers, A. (2016a). Modelling as indirect representation? The Lotka– Volterra Model revisited. *British Journal for the Philosophy of Science*, axv055-30. https://doi.org/10.1093/bjps/axv055.

Knuuttila, T., & Loettgers, A. (2016b). Model templates within and between disciplines: From magnets to gases – and socio-economic systems. *European Journal for Philosophy of Science*, *6*(3), 377–400. https://doi.org/10.1007/s13194-016-0145-1.

Koricheva, J., Gurevitch, J., & Mengersen, K. (Eds.). (2013). *Handbook of Meta-analysis in Ecology and Evolution*. Princeton University Press.

Kulmatiski, A., Heavilin, J., & Beard, K. H. (2011). Testing predictions of a three-species plant-soil feedback model. *Journal of Ecology*, *99*(2), 542–50.

Ladyman, J., Lambert, J., & Wiesner, K. (2013). What is a complex system? *European Journal for Philosophy of Science*, *3*, 33–67.

Lange, M. (2005). Ecological laws: What would they be and why would they matter? *Oikos*, *110*(2), 394–403.

Lawton, J. H. (1999). Are there general laws in ecology? *Oikos*, *84*(2), 177–92.

Levins, R. (1993). A response to Orzack and Sober: Formal analysis and the fluidity of science. *The Quarterly Review of Biology*, *68*(4), 547–55. https://doi.org/10.1086/418302.

Levin, S. A. (1998). Ecosystems and the biosphere as complex adaptive systems. *Ecosystems*, *1*(5), 431–6.

Levin, S. A. (2002). Complex adaptive systems: Exploring the known, the unknown and the unknowable. *Bulletin of the American Mathematical Society*, *40*(1), 3–19.

Levin, S. A. (2005). Self-organization and the emergence of complexity in ecological systems. *BioScience, 55*(12), 1075–9.

Levins, R. (1966). The strategy of model building in population biology. *American Scientist, 54*(4), 421–31.

Levy, A. (2018). Idealization and abstraction: Refining the distinction. *Synthese, 13*(1), 1–18.

Linquist, S., T. R. Gregory, T. A. Elliott, et al. 2016. "Yes! there are resilient generalizations (or 'laws') in ecology." *Quarterly Review of Biology, 91*(2): 119–31. http://doi.org/10.1086/686809.

Lipton, P. (2005). Testing hypotheses: Prediction and prejudice. *Science, 307*(5707), 219–21.

Lockwood, J. L., Cassey, P., & Blackburn, T. (2005). The role of propagule pressure in explaining species invasions. *Trends in Ecology & Evolution, 20*(5), 223–8.

Loreau, M., Naeem, S., Inchausti, P., et al. (2001). Biodiversity and ecosystem functioning: current knowledge and future challenges. *Science, 294*(5543), 804–8.

Maclaurin, J., & Sterelny, K. (2008). *What Is Biodiversity?* University of Chicago Press.

Marquet, P. A., Allen, A. P., Brown, J. H. et al. (2014). On theory in ecology. *BioScience, 64*(8), 701–10.

Marshall, K. E., & Sinclair, B. J. (2012). Threshold temperatures mediate the impact of reduced snow cover on overwintering freeze-tolerant caterpillars. *Naturwissenschaften, 99*(1), 33–41.

Matthewson, J. (2011). Trade-offs in model-building: A more target-oriented approach. *Studies in History and Philosophy of Science Part A, 42*(2), 324–33.

May, R.M. (1973). *Stability and Complexity in Model Ecosystems.* Princeton University Press.

McCann, K. S. (2000). The diversity-stability debate. *Nature, 405*(6783), 228–33. https://doi.org/10.1038/35012234.

McIntosh, R. P. (1987). Pluralism in ecology. *Annual Review of Ecology and Systematics, 18*(1), 321–41.

McShea, D. W., & Brandon, R. N. (2010). *Biology's First Law.* University of Chicago Press.

Miller, J. H., & Page, S. E. (2009). *Complex Adaptive Systems.* Princeton University Press.

Mills, L. S., & Doak, D. F. (1993). The keystone-species concept in ecology and conservation. *BioScience, 43*(4), 219–24.

Mitchell, Sandra D. (2000). Dimensions of scientific law. Philosophy of Science 67 (2):242–265.

Mitchell, S. D. (2003). *Biological Complexity and Integrative Pluralism*. Cambridge University Press.

Mitchell, S. D. (2009). *Unsimple Truths*. University of Chicago Press.

Morgan, M. S. (2005). Experiments versus models: New phenomena, inference and surprise. *Journal of Economic Methodology, 12*(2), 317–29. http://doi.org/10.1080/13501780500086313.

Morgan, M. S., & Morisson, M. (1999). *Models as Mediators*. Cambridge University Press.

Novak, M., Wootton, J. T., Doak, D. F. et al. (2011). Predicting community responses to perturbations in the face of imperfect knowledge and network complexity. *Ecology, 92*(4), 836–46.

Odenbaugh, J. (2003). Complex systems, trade-offs, and theoretical population biology: Richard Levins's 'strategy of model building in population biology' revisited. *Philosophy of Science (Proceedings), 70*(5), 1496–507.

Odenbaugh, J. (2011). *Complex Ecological Systems* (C. Hooker, Ed., Vol. 10, pp. 421–431). Elsevier.

Orzack, S. H. (2005). Discussion: What, if anything, is 'The strategy of model building in population biology?' A comment on Levins (1966) and Odenbaugh (2003). *Philosophy of Science, 72*(3), 479–85.

Orzack, S. H., & Sober, E. (1993). A critical assessment of Levins's the strategy of model building in population biology (1966). *The Quarterly Review of Biology, 68*(4), 533–46.

Parke, E. (2014). Experiments, simulations, and epistemic privilege. *Philosophy of Science, 81*(4), 516–36.

Parker, J., Burkepile, D. E., & Hay, M. E. (2006). Opposing effects of native and exotic herbivores on plant invasions. *Science, 311*(5766), 1459–61.

Parker, W. (2010). Predicting weather and climate: Uncertainty, ensembles and probability. *Studies in History and Philosophy of Science Part B: Studies in History and Philosophy of Modern Physics, 41*(3), 263–72.

Parker, W., & Risbey, J. S. (2015). False precision, surprise and improved uncertainty assessment. *Philosophical Transactions of the Royal Society of London A: Mathematical, Physical and Engineering Sciences, 373*(2055), 20140453.

Parrish, J. K., & Edelstein-Keshet, L. (1999). Complexity, pattern, and evolutionary trade-offs in animal aggregation. *Science, 284*(5411), 99–101.

Parrott, L. (2010). Measuring ecological complexity. *Ecological Indicators, 10*(6), 1069–76.

Peters, D., B. Bestelmeyer, & J Herrick. 2006. Disentangling complex landscapes: New insights into arid and semiarid system dynamics. *BioScience, 56*(6), 491–501.

Peters, R. H. (1991). *A Critique for Ecology*. Cambridge University Press.

Phillips, R. P., Ibanez, I., & D'Orangeville, L. (2016). A belowground perspective on the drought sensitivity of forests: Towards improved understanding and simulation. *Forest Ecology and Management*, 380, 309–20.

Pine, W. E., Pollock, K. H., Hightower, J. E., Kwak, T. J., & Rice, J. A. (2003). A review of tagging methods for estimating fish population size and components of mortality. *Fisheries*, 28(10), 10–23.

Potochnik, A. (2017). *Idealization and the Aims of Science*. University of Chicago Press.

Proctor, J. D., & Larson, M. (2005). Ecology, complexity, and metaphor. *BioScience*, 55(12), 1065–8.

Raerinne, J. (2011). Causal and mechanistic explanations in ecology. *Acta Biotheoretica*, 59 251–71.

Ramsey, D. S. L., Forsyth, D. M., Veltman, C. J. et al. (2012). An approximate Bayesian algorithm for training fuzzy cognitive map models of forest responses to deer control in a New Zealand adaptive management experiment. *Ecological Modelling*, 240, 93–104.

Ramsey, D. S. L., & Veltman, C. J. (2005). Predicting the effects of perturbations on ecological communities: What can qualitative models offer? *Journal of Animal Ecology*, 74(5), 905–16.

Ricklefs, R. E., & Miller, G. L. (2000). *Ecology*. Freeman and Company. *Recovery Strategy for the Vancouver Island Marmot (Marmota vancouverensis) in British Columbia*. Prepared for the B.C. Ministry of Environment, Victoria, BC. 25 pp.

Rind, D. (1999). Complexity and climate. *Science*, 284(5411), 105–7.

Rosenberg, A. (1989). Are generic predictions enough? *Philosophy of Economics*: Proceedings, Munich, July 1981 (Vol. 2). Springer Science & Business Media.

Salmon, M. H., Earman, J., Glymour, C., & Lennox, J. G. (1992). *Introduction to the Philosophy of Science*. Hackett.

Salmon, W. C. (2006). *Four Decades of Scientific Explanation*. University of Pittsburgh Press.

Santana, C. (2014). Save the planet: Eliminate biodiversity. *Biology & Philosophy*, 29(6), 761–80.

Scerri, E. R. (2006). *The Periodic Table: Its Story and Its Significance*. Oxford University Press.

Scerri, E. R., & Worrall, J. (2001). Prediction and the periodic table. *Studies in History and Philosophy of Science Part A*, 32(3), 407–52.

Schindler, D. E., & Hilborn, R. (2015). Prediction, precaution, and policy under global change. *Science*, 347(6225), 953–4.

Shrader-Frechette, K. S. & McCoy E. D. (1993). Method in ecology: strategies for conservation. New York, NY, USA: Cambridge University Press. Edited by Earl D. McCoy.

Simon, H. A. (1962). The architecture of complexity. *Proceedings of the American Philosophical Society, 106*(6), 467–82.

Sinclair, B. J., Jako Klok, C., Scott, M. B., Terblanche, J. S., & Chown, S. L. (2003). Diurnal variation in supercooling points of three species of Collembola from Cape Hallett, Antarctica. *Journal of Insect Physiology, 49*(11), 1049–61.

Sinclair, B. J., Vernon, P., Jaco Klok, C., & Chown, S. L. (2003). Insects at low temperatures: An ecological perspective. *Trends in Ecology & Evolution, 18*(5), 257–62.

Singer, M. C., & Parmesan, C. (2018). Lethal trap created by adaptive evolutionary response to an exotic resource. *Nature, 557*(7704), 238–41.

Smith, L. A., & Stern, N. (2011). Uncertainty in science and its role in climate policy. *Philosophical Transactions of the Royal Society of London A: Mathematical, Physical and Engineering Sciences, 369*(1956), 4818–41.

Sobek-Swant, S., Kluza, D. A., Cuddington, K., & Lyons, D. B. (2012). Potential distribution of emerald ash borer: What can we learn from ecological niche models using Maxent and GARP? *Forest Ecology and Management, 281*, 23–31.

Sober, E. (2011). A priori causal models of natural selection. *Australasian Journal of Philosophy, 89*(4), 571–89.

Stegenga, J. (2011). Is meta-analysis the platinum standard of evidence? *Studies in History and Philosophy of Biological and Biomedical Sciences, 42*(4), 497–507.

Stenseth, N. C. (1999). Population cycles in voles and lemmings: Density dependence and phase dependence in a Stochastic world. *Oikos, 87*(3), 427–61.

Stillman, R. A., Railsback, S. F., Giske, J., Berger, U., & Grimm, V. (2015). Making predictions in a changing world: The benefits of individual-based ecology. *BioScience, 65*(2), 140–50.

Stockwell, D. (1999). The GARP modelling system: Problems and solutions to automated spatial prediction. *International Journal of Geographical Information Science, 13*(2), 143–58.

Storch, D., & Gaston, K. J. (2004). Untangling ecological complexity on different scales of space and time. *Basic and Applied Ecology, 5*(5), 389–400.

Strevens, M. (2004). The causal and unification approaches to explanation unified – causally. *Noûs, 38*(1), 154–76.

Suding, K. N., Stanley Harpole, W., Fukami, T. et al. (2013). Consequences of plant-soil feedbacks in invasion. *Journal of Ecology, 101*(2), 298–308.

Tang, Y., Wu, Y., Liu, K. et al. (2019). Investigating the distribution of the Yangtze finless porpoise in the Yangtze River using environmental DNA. *PLoS One, 14*(8).

Têmkin, I. (2021) Phenomenological Levels in Biological and Cultural Evolution in Brooks, D. S., DiFrisco, J., & Wimsatt, W. C. (Eds.). (2021). *Levels of organization in the biological sciences*. MIT Press. 297–316

Thiengo, S. C., Faraco, F. A., Salgado, N. C., Cowie, R. H., & Fernandez, M. A. (2007). Rapid spread of an invasive snail in South America: The giant African snail, Achatina fulica, in Brasil. *Biological Invasions, 9*(6), 693–702.

Tompkins, D. M., & Veltman, C. J. (2006). Unexpected consequences of vertebrate pest control: Predictions from a four-species community model. *Ecological Applications: A Publication of the Ecological Society of America, 16*(3), 1050–61.

Travis, J., Coleman, F. C., & Auster, P. J. (2014). Integrating the invisible fabric of nature into fisheries management. *Proceedings of the National Academy of Sciences of the United States of America, 111*(2), 581–4.

Turchin, P. (2001). Does population ecology have general laws? *Oikos, 94*(1), 17–26.

Tyne, J. A., Loneragan, N. R., Johnston, D. W. et al. (2016). Evaluating monitoring methods for cetaceans. *Biological Conservation, 201*, 252–60.

USDA Forest Service. (2020). *Emerald Ash Borer*. Emerald Ash Borer Information Network. www.emeraldashborer.info/.

Valéry, L., Fritz, H., & Lefeuvre, J. C. (2013). Another call for the end of invasion biology. *Oikos, 122*(8), 1143–6.

Valls, A., Coll, M., & Christensen, V. (2015). Keystone species: Toward an operational concept for marine biodiversity conservation. *Ecological Monographs, 85*(1), 29–47.

van der Putten, W. H., Bardgett, R. D., Bever, J. D. et al. (2013). Plant-soil feedbacks: The past, the present and future challenges. *Journal of Ecology, 101*(2), 265–76.

Van Fraassen, B. C. (2008). *Scientific Representation: Paradoxes of Perspective* (Vol. 70). Oxford University Press.

Wang, L., & Jackson, D. A. (2014). Shaping up model transferability and generality of species distribution modeling for predicting invasions: Implications from a study on Bythotrephes longimanus. *Biological Invasions, 16*(10), 2079–103.

Ward, E. J., Holmes, E. E., Thorson, J. T., & Collen, B. (2014). Complexity is costly: A meta-analysis of parametric and non-parametric methods for short-term population forecasting. *Oikos, 123*(6), 652–61.

Weisberg, M. (2004). Qualitative theory and chemical explanation. *Philosophy of Science, 71*(5), 1071–81.

Weisberg, M. (2006). Forty years of 'the strategy': Levins on model building and idealization. *Biology & Philosophy, 21*, 623–45.

Weisberg, M. (2007). Three kinds of idealization. *The Journal of Philosophy*, 639–659.

Weisberg, M. (2013). *Simulation and Similarity*. Oxford University Press USA.

Weng, G., Bhalla, U. S., & Iyengar, R. (1999). Complexity in biological signaling systems. *Science, 284*(5411), 92–6.

Wenger, S. J., & Olden, J. D. (2012). Assessing transferability of ecological models: An underappreciated aspect of statistical validation. *Methods in Ecology and Evolution, 3*(2), 260–7.

Whitesides, G. M., & Ismagilov, R. F. (1999). Complexity in chemistry. *Science, 284*(5411), 89–92.

Wilcox, C. (2018, 27 August). When snails attack: The epic discovery of an ecological Phenomenon. *Discover.* www.discovermagazine.com/planet-earth/when-snails-attack-the-epic-discovery-of-an-ecological-phenomenon.

Wimsatt, W. C. (1972). Complexity and organization. *Proceedings of the Biennial Meeting of the Philosophy of Science Association, 1972*, 67–86.

Winther, R. G. (2011). Prediction in selectionist evolutionary theory. *Philosophy of Science, 76*(5), 889–901.

Woodward, J. (2001). Law and explanation in biology: Invariance is the kind of stability that matters. *Philosophy of Science, 68*(1), 1–20.

Woodward, J. (2010). Causation in biology: Stability, specificity, and the choice of levels of explanation. *Biology & Philosophy, 25*(3), 287–318.

Acknowledgements

I am grateful to Adrian Currie, Angela Potochnik, the members of TINT at University of Helsinki, the Philosophy of Biology group at University of Bielefeld, especially Marie Kaiser, and two referees for helpful and constructive feedback on drafts of the manuscript. This work would not have been achieved without the help of Kim Sterelny, who motivated me to start this project, and Brent Sinclair, whose signature phrase "except when they don't" triggered my interest in ecological complexity. Thanks also to Grant Ramsey and Michael Ruse. Much of the research for this Element was funded by H2020 Marie Skłodowska-Curie Actions (Grant No. 796910). I am indebted to my mother, Lydia, for feedback on how to write legibly. Last but not least, I am eternally grateful to Yiannis Kokosalakis for all his support through thick and thin.

Dedicated to my parents, for believing in me and encouraging all my academic endeavours.

Cambridge Elements ☰

Philosophy of Biology

Grant Ramsey

KU Leuven, Belgium

Grant Ramsey is a BOFZAP research professor at the Institute of Philosophy, KU Leuven, Belgium. His work centers on philosophical problems at the foundation of evolutionary biology. He has been awarded the Popper Prize twice for his work in this area. He also publishes in the philosophy of animal behavior, human nature and the moral emotions. He runs the Ramsey Lab (theramseylab.org), a highly collaborative research group focused on issues in the philosophy of the life sciences.

Michael Ruse

Florida State University

Michael Ruse is the Lucyle T. Werkmeister Professor of Philosophy and the Director of the Program in the History and Philosophy of Science at Florida State University. He is Professor Emeritus at the University of Guelph, in Ontario, Canada. He is a former Guggenheim fellow and Gifford lecturer. He is the author or editor of over sixty books, most recently *Darwinism as Religion: What Literature Tells Us about Evolution*; *On Purpose*; *The Problem of War: Darwinism, Christianity, and their Battle to Understand Human Conflict*; and *A Meaning to Life*.

About the Series

This Cambridge Elements series provides concise and structured introductions to all of the central topics in the philosophy of biology. Contributors to the series are cutting-edge researchers who offer balanced, comprehensive coverage of multiple perspectives, while also developing new ideas and arguments from a unique viewpoint.

Cambridge Elements ☰

Philosophy of Biology

Elements in the Series

Adaptation
Elisabeth A. Lloyd

Stem Cells
Melinda Bonnie Fagan

The Metaphysics of Biology
John Dupré

Facts, Conventions, and the Levels of Selection
Pierrick Bourrat

The Causal Structure of Natural Selection
Charles H. Pence

Philosophy of Developmental Biology
Marcel Weber

Evolution, Morality and the Fabric of Society
R. Paul Thompson

Structure and Function
Rose Novick

Hylomorphism
William M. R. Simpson

Biological Individuality
Alison K. McConwell

Human Nature
Grant Ramsey

Ecological Complexity
Alkistis Elliott-Graves

A full series listing is available at www.cambridge.org/EPBY

Printed in the United States
by Baker & Taylor Publisher Services

Printed in the United States
by Baker & Taylor Publisher Services